LONDON MATHEMATICAL SOCIETY STUDENT TEXTS

Managing editor: Dr C.M. Series, Mathematics Institute,
University of Warwick, Coventry CV4 7AL, United Kingdom

London Mathematical Society Student Texts 23

Complex Algebraic Curves

Frances Kirwan

Mathematical Institute, University of Oxford

CAMBRIDGE
UNIVERSITY PRESS

CAMBRIDGE UNIVERSITY PRESS
Cambridge, New York, Melbourne, Madrid, Cape Town, Singapore,
São Paulo, Delhi, Dubai, Tokyo, Mexico City

Cambridge University Press
The Edinburgh Building, Cambridge CB2 8RU, UK

Published in the United States of America by
Cambridge University Press, New York

www.cambridge.org
Information on this title: www.cambridge.org/9780521423533

© Cambridge University Press 1992

First published 1992
Reprinted 1993, 1995

A catalogue record for this publication is available from the British Library

Library of Congress Cataloguing in Publication Data

ISBN 978-0-521-41251-3 Hardback
ISBN 978-0-521-42353-3 Paperback

Contents

Preface

This book on complex algebraic curves is intended to be accessible to any third year mathematics undergraduate who has attended courses on algebra, topology and complex analysis. It is an expanded version of notes written to accompany a lecture course given to third year undergraduates at Oxford. It has usually been the case that a number of graduate students have also attended the course, and the lecture notes have been extended somewhat for the sake of others in their position. However this new material is not intended to daunt undergraduates, who can safely ignore it. The original lecture course consisted of Chapters 1 to 5 (except for some of §3.1 including the definition of intersection multiplicities) and part of Chapter 6, although some of the contents of these chapters (particularly the introductory material in Chapter 1) was covered rather briefly.

Each section of each chapter has been arranged as far as possible so that the important ideas and results appear near the start and the more difficult and technical proofs are left to the end. Thus there is no need to finish each section before beginning the next; when the going gets tough the reader can afford to skip to the start of the next section.

The main aim of the course was to show undergraduates in their final year how the basic ideas of pure mathematics they had studied in previous years could be brought together in one of the showpieces of mathematics. In particular it was intended to provide those students not intending to continue mathematics beyond a first degree with a final year course which could be regarded as a culmination of their studies, rather than one consisting of the development of more machinery which they would never have the opportunity to use. As well as being one of the most beautiful areas of mathematics, the study of complex algebraic curves is one in which it is not necessary to develop new machinery before starting - the tools are already available from basic algebra, topology and complex analysis. It was also hoped that the course would give those students who might be tempted to continue mathematics an idea of the flavour, or rather the very varied and exciting array of flavours, of algebraic geometry, illustrating the way it draws on all parts of mathematics while avoiding as much as possible the elaborate and highly developed technical foundations of the subject.

The contents of the book are as follows. Chapter 1 "can be omitted for examination purposes" as the original lecture notes said. This chapter is simply intended to provide some motivation and historical background for the study of complex algebraic curves, and to indicate a few of the numerous reasons why they are of interest to mathematicians working in very different areas. Chapter 2 lays the foundations with the technical definitions and basic results needed to start the subject. Chapter 3 studies algebraic questions about complex algebraic curves, in particular the question of how two curves meet each other. Chapter 4 investigates what complex algebraic curves look like topologically. In Chapters 5 and 6 complex analysis is used to investigate complex algebraic curves from a third point of view. Finally Chapter 7 looks at *singular* complex algebraic curves which are much more complicated objects than nonsingular ones and are mostly ignored in the first six chapters. The three appendices contain results from algebra, complex analysis and topology which are included to make the book as self- contained as possible: they are not intended to be easily readable but simply to be available for those who feel the need to consult them.

There are many excellent books available for those who wish to study the subject further: see for example the books by Arbarello & al, Beardon, Brieskorn and Knörrer, Chern, Clemens, Coolidge, Farkas and Kra, Fulton, Griffiths, Griffiths and Harris, Gunning, Hartshorne, Jones, Kendig, Morrow and Kodaira, Mumford, Reid, Semple and Roth, Shafarevich, Springer, and Walker listed in the bibliography. Many of these references I have used to prepare the lecture course and accompanying notes on which this book was based, as well as the book itself. Indeed, the only reason I had for writing lecture notes and then this book was that each of the books listed either assumes a good deal more background knowledge than undergraduates are likely to have or else takes a very different approach to the subject.

Finally I would like to record my grateful thanks to Graeme Segal, for first suggesting that an undergraduate lecture course on this subject would be worthwhile, to all those students who attended the lecture course and the graduate students who helped run the accompanying classes for their useful comments, to David Tranah of the Cambridge University Press and Elmer Rees for their encouragement and advice on turning the lecture notes into a book, and to Mark Lenssen and Amit Badiani for their great help in producing the final version.

Frances Kirwan
Balliol College, Oxford
August 1991

Chapter 1

Introduction and background

A complex algebraic curve in $\mathbf{C}^2$ is a subset C of $\mathbf{C}^2 = \mathbf{C} \times \mathbf{C}$ of the form

$$C = \{(x,y) \in \mathbf{C}^2 : P(x,y) = 0\} \qquad (1.1)$$

where $P(x,y)$ is a polynomial in two variables with complex coefficients. (See §2.1 for the precise definition). Such objects are called curves by analogy with real algebraic curves or "curved lines" which are subsets of $\mathbf{R}^2$ of the form

$$\{(x,y) \in \mathbf{R}^2 : P(x,y) = 0\} \qquad (1.2)$$

where $P(x,y)$ is now a polynomial with real coefficients.

Of course to each real algebraic curve there is associated a complex algebraic curve defined by the same polynomial. Real algebraic curves were studied long before complex numbers were recognised as acceptable mathematical objects, but once complex algebraic curves appeared on the scene it quickly became clear that they have at once simpler and more interesting properties than real algebraic curves. To get some idea why this should be, consider the study of polynomial equations in one variable with real coefficients: it is easier to work with complex numbers, so that the polynomial factorises completely, and then decide which roots are real than not to allow the use of complex numbers at all.

In this book we shall study complex algebraic curves from three different points of view: algebra, topology and complex analysis. An example of the kind of algebraic question we shall ask is

"Do the polynomial equations
$$P(x,y) = 0$$
and
$$Q(x,y) = 0$$
defining two complex algebraic curves have any common solutions $(x,y) \in \mathbf{C}^2$, and if so, how many are there?"

An answer to this question will be given in Chapter 3.

The relationship of the study of complex algebraic curves with complex analysis arises when one attempts to make sense of "multi-valued holomorphic functions" such as

$$z \mapsto z^{\frac{3}{2}}$$

and

$$z \mapsto (z^3 + z^2 + 1)^{\frac{1}{2}}.$$

One ends up looking at the corresponding complex algebraic curves, in these cases

$$y^2 = x^3$$

and

$$y^2 = x^3 + x^2 + 1.$$

Complex analysis will be important in Chapter 5 and Chapter 6 of this book.

We shall also investigate the topology (that is, roughly speaking, the shape) of complex algebraic curves in Chapter 4 and §7.3. It is of course not possible to sketch a complex algebraic curve in $\mathbf{C}^2$ in the same way that we can sketch real algebraic curves in $\mathbf{R}^2$, because $\mathbf{C}^2$ has four real dimensions. None the less, we can draw sketches of complex algebraic curves (with some extra points added "at infinity"), which are accurate *topological* pictures of the curves but which do not reflect the way they sit inside $\mathbf{C}^2$. For some examples, see figure 1.1. It is important to stress the fact that these pictures can only represent the complex curves as topological spaces, and not the way they lie in $\mathbf{C}^2$. For example, the complex curve defined by $xy = 0$ is the union of the two "complex lines" defined by $x = 0$ and $y = 0$ in $\mathbf{C}^2$, which meet at the origin $(0,0)$. Topologically when we add a point at infinity to each complex line it becomes a sphere, and the complex curve becomes the union of two spheres meeting at a point as in figure 1.2. This picture, though topologically correct, represents the two complex lines as tangential to each other at the point of intersection, and this is not the case in $\mathbf{C}^2$. We cannot avoid this problem without making the complex lines look "singular" at the origin as in figure 1.3, which again is not really the case.

1.1   A brief history of algebraic curves.

Real algebraic curves have been studied for more than two thousand years, although it was not until the introduction of the systematic use of coordinates into geometry in the seventeenth century that they could be described in the form (1.2).

| Equation | Real algebraic curve | Complex algebraic curve (with points "at infinity") |

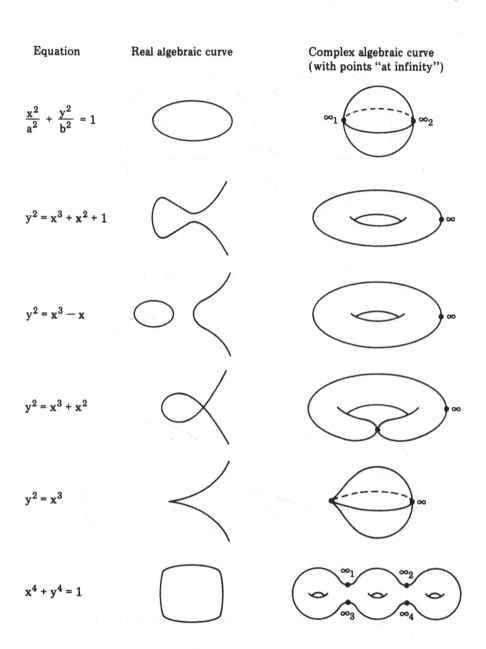

Figure 1.1: Some algebraic curves

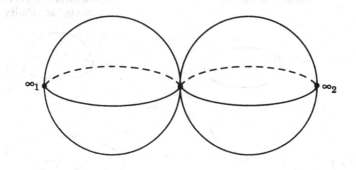

Figure 1.2: The complex curve $xy = 0$

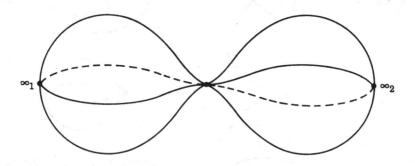

Figure 1.3: Another view of the complex curve $xy = 0$

The story starts with the Greeks, who had very sophisticated geometrical methods but a relatively primitive understanding of algebra. To them a circle was not defined by an equation

$$(x-a)^2 + (y-b)^2 = r^2$$

but was instead the locus of all points having equal distance r from a fixed point $P = (a, b)$. Similarly a parabola to the Greeks was the locus of all points having equal distance from a given point P and a given line L, while an ellipse (hyperbola) was the locus of all points for which the sum (difference) of the distances from two given points P and Q had a fixed value.

Lines and circles can of course be drawn with a ruler and compasses, and the Greeks devised more complicated mechanisms to construct parabolas, ellipses and hyperbolas. With these they were able to solve some famous problems such as "duplicating the cube"; in other words constructing a cube whose volume is twice the volume of a given cube (this was called the Delian problem. This comes down to constructing a line segment of length $2^{1/3}$ times the length of a given unit segment. The Greeks realised that this could be done by constructing the points of intersection of the parabolas

$$y^2 = 2x$$

and

$$x^2 = y.$$

They tried very hard to do this and other constructions (such as trisecting an arbitrary angle and drawing regular polygons) using ruler and compasses alone. They failed, and in fact it can be shown using Galois theory (see for example [Stewart 73] pp.57-67) that these constructions are impossible with ruler and compasses.

Besides lines and circles, ellipses, parabolas and hyperbolas the Greeks knew constructions for many other curves, for example the epicyclic curves used to describe the paths of planets before the discovery of Kepler's laws. (An epicyclic curve is the path of a point on a circle which rolls without slipping on the exterior of a fixed circle: see figure 1.4). Greek mathematics was almost forgotten in Western Europe for many centuries after the end of the Roman Empire, but in the late Middle Ages and Renaissance period it was gradually rediscovered through contact with Arab mathematicians. It was during the Renaissance that new algebraic curves were discovered by artists such as Leonardo da Vinci who were interested in drawing outlines of three-dimensional shapes in perspective.

As well as reintroducing Greek mathematics the Arabs introduced to Europe a much more sophisticated understanding of algebra and a good algebraic

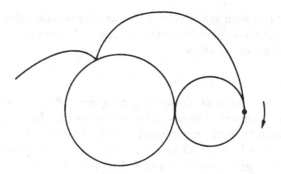

Figure 1.4: An epicyclic curve

notation. It can be difficult for us to realise how important good notation is in the solution of a mathematical problem. For example the simple argument

$$x^2 + 3 = 5x \Rightarrow (x - 5/2)^2 = 13/4 \Rightarrow x = (5 \pm \sqrt{13})/2$$

becomes much harder to express and to follow using words alone.

By the end of the seventeenth century mathematicians were familiar with the idea pioneered by Descartes and Fermat of describing a locus of points in the plane by one or more equations in two variables x and y. The methods of the differential calculus were gradually being understood and applied to curves. It was known that many real algebraic curves turned up in problems in applied mathematics (one example being the nephroid or kidney curve which is seen when light is reflected from a mirror whose cross-section is part of a circle).

Around 1700 Newton made a detailed study of cubic curves (that is, curves defined by polynomials of degree three) and described seventy two different cases. He investigated the *singularities* of a curve C defined by a polynomial $P(x,y)$, i.e. the points $(x,y) \in C$ satisfying

$$\frac{\partial P}{\partial x}(x,y) = 0 = \frac{\partial P}{\partial y}(x,y).$$

These are points where the curve does not look "smooth", such as the origin in the cubic curves defined by $y^2 = x^3 + x^2$ and $y^2 = x^3$ (cf. figure 1.1). We shall investigate the singularities of curves in greater detail in Chapter 7.

Once the use of complex numbers was understood in the nineteenth century it was realised that very often it is easier and more profitable to study the complex solutions to a polynomial equation $P(x,y) = 0$ instead of just the real solutions. For example, if we allow complex projective changes of coordinates

$$(x,y) \mapsto \left(\frac{ax + by + c}{hx + jy + k}, \frac{dx + ey + f}{hx + jy + k} \right)$$

where

$$\begin{pmatrix} a & b & c \\ d & e & f \\ h & j & k \end{pmatrix}$$

is a nonsingular matrix (see Chapter 2 for more details), then many of Newton's seventy two different cubics become equivalent to one another. In fact any complex curve defined by an irreducible cubic polynomial can be put into one of the forms

$$\begin{aligned} y^2 &= x(x-1)(x-\lambda) && \text{with } \lambda \neq 0,1 \text{ (nonsingular cubic)} \\ y^2 &= x^2(x+1) && \text{(nodal cubic)} \\ y^2 &= x^3 && \text{(cuspidal cubic)} \end{aligned}$$

(see corollary 3.34 and exercise 3.9).

Another example of the "better" behaviour of complex curves than real ones is the fact that a real algebraic curve can be so degenerate it doesn't look like a curve at all. For example, the subset

$$\{(x,y) \in \mathbf{R}^2 : x^2 + y^2 = 0\}$$

of $\mathbf{R}^2$ is the single point $(0,0)$ and

$$\{(x,y) \in \mathbf{R}^2 : x^2 + y^2 = -1\}$$

is the empty set. But if $P(x,y)$ is any nonconstant polynomial with complex coefficients then the subset of the *complex* space $\mathbf{C}^2$ given by

$$\{(x,y) \in \mathbf{C}^2 : P(x,y) = 0\}$$

is nonempty and "has complex dimension one" in a reasonable sense.

In the nineteenth century it was realised that if suitable "points at infinity" are added to a complex algebraic curve it becomes a compact topological space, just as the Riemann sphere is made by adding an extra point ∞ to $\mathbf{C}$. Moreover one can make sense of the concepts of holomorphic and meromorphic functions on this topological space, and much of the theory of complex analysis on $\mathbf{C}$ can be applied. This leads to the theory of Riemann surfaces[1],

---

[1] There is an unfortunate inconsistency of terminology in the theory of complex algebraic curves and Riemann surfaces. A complex algebraic curve is called a curve because its *complex* dimension is one, but its *real* dimension is two so it can also be called a surface.

called after Bernhard Riemann (1826-1866). Riemann was extremely influential in developing the idea that geometry should deal not only with ordinary Euclidean space but also with much more general and abstract spaces.

At the same time as Riemann and his followers were investigating complex algebraic curves using complex analysis and topology, other mathematicians began to use purely algebraic methods to obtain the same results. In 1882 Dedekind and Weber showed that much of the theory of algebraic curves remained valid when the field of complex numbers was replaced by another (preferably algebraically closed) field K. Instead of studying the curve C defined by an irreducible polynomial $P(x,y)$ directly, they studied the "field of rational functions on C" which consists of all functions $f : C \rightarrow K \cup \{\infty\}$ of the form

$$f(x,y) = Q(x,y)/R(x,y)$$

where $Q(x,y)$ and $R(x,y)$ are polynomials with coefficients in K such that $R(x,y)$ is not divisible by $P(x,y)$ (i.e. such that $R(x,y)$ does not vanish identically on C).

It is often useful to study curves defined over fields other than the fields of real and complex numbers. For example number theorists interested in the integer solutions to a diophantine equation

$$P(x,y) = 0,$$

where $P(x,y)$ is a polynomial with integer coefficients, often first regard the equation as a congruence modulo a prime number p. Such a congruence can be thought of as defining an algebraic curve over the finite field

$$\mathbf{F}_p = \mathbf{Z}/p\mathbf{Z}$$

consisting of the integers modulo p, or over its algebraic closure.

By the end of the nineteenth century mathematicians had begun to make progress in studying the solutions of systems of more than one polynomial equation in more than two variables. During the twentieth century many more ideas and results have been developed in this area of mathematics (known as algebraic geometry). Algebraic curves and surfaces are now reasonably well understood, but the theory of algebraic varieties of dimension greater than two remains very incomplete. (An algebraic variety is, roughly speaking, the set of solutions to finitely many polynomial equations in finitely many variables over a field K).

The study of algebraic curves and Riemann surfaces, involving as it does a rich interplay between algebra, analysis, topology and geometry, with applications in many different areas of mathematics, has been the subject of active research right up to the present day. Just one example of its importance in modern research is a theory of the structure of elementary physical particles which received much attention in the 1980s. In this theory particles are represented by "strings" or loops, and the surfaces in space-time swept out by these strings as they develop from creation to annihilation are Riemann surfaces which are closely related to complex algebraic curves in the complex plane C^2 with some points added at infinity.

1.2 Relationship with other parts of mathematics

Nowadays complex algebraic curves turn up and are useful in all sorts of areas of mathematics ranging from theoretical physics to number theory. This section gives a brief explanation of their importance in a few of these areas.

1.2.1 Number theory

In the rest of this book the links between the study of complex algebraic curves and number theory will not be stressed, but they are important.

Number theorists are interested in the integer solutions to equations such as

$$x^n + y^n = z^n. \tag{1.3}$$

For example, does this equation have any solutions with x, y, and z nonzero integers when $n > 2$? This comes down to the question of whether there exist nonzero rational solutions

$$s = x/z, t = y/z,$$

to the equation

$$s^n + t^n = 1, \tag{1.4}$$

which defines a complex algebraic curve called the Fermat curve of degree n. This curve is called after the French mathematician Pierre Fermat (1601-1665) because Fermat wrote in the margin of one of his books that he had found a "truly marvellous proof" (*demonstrationem mirabilem sane*) that the equation (1.3) has no nonzero integer solutions when $n > 2$. For more than three hundred years mathematicians have been trying to prove or disprove this statement, without success although it has been proved that there are no nonzero integer solutions for many particular values of n. In 1983 the German mathematician Faltings proved that a complex algebraic curve of genus at least two has only finitely many points with rational coefficients (see Chapter 4 for the definition of genus). The Fermat curve of degree n has genus

$$\frac{1}{2}(n-1)(n-2)$$

(see §4.3) so when $n \geq 4$ it follows that there are only a finite number of rational solutions to (1.4). Whether that finite number is ever nonzero, nobody knows.

1.2.2   Singularities and the theory of knots

Another important area which will just be touched on in this book is the study of singularities (see Chapter 7). For more details and lots of pictures see [Brieskorn & Knörrer 86].

A singularity of an algebraic curve C defined by a polynomial equation

$$P(x,y) = 0$$

is a point $(a,b) \in \mathbf{C}$ satisfying

$$\frac{\partial P}{\partial x}(a,b) = 0 = \frac{\partial P}{\partial y}(a,b).$$

For example the curve

$$y^2 = x^3 + x^2 + 1$$

has no singularities, whereas the curves

$$y^2 = x^3 + x^2$$

and

$$y^2 = x^3$$

have singularities at the origin. Figure 1.1 showed the real points on these curves.

We have already observed that drawing the real points on a complex algebraic curve is not always helpful in studying the complex curve (e.g. for the curve $x^2 + y^2 + 1 = 0!$) though it can give some idea of what is going on near a singularity with real coordinates. A better way to study what the curve looks like near a singularity (a,b) is to look at its intersection with a small three-dimensional sphere

$$\{(x,y) \in \mathbf{C}^2 : |\, x - a\,|^2 + |\, y - b\,|^2 = \varepsilon^2\}$$

in $\mathbf{C}^2 = \mathbf{R}^4$ with centre at (a, b). This intersection will be a "knot" or "link" in the three-dimensional sphere. We can identify this three-dimensional sphere topologically with Euclidean space $\mathbf{R}^3$ together with an extra point at infinity by using stereographic projection. Let us assume for simplicity that the centre (a, b) of the sphere is the origin $(0, 0)$ in $\mathbf{C}^2$. Then stereographic projection of the sphere from the point $(0, \varepsilon)$ maps each point (x, y) of the sphere *other* than $(0, \varepsilon)$ to the point of intersection of the (real) line joining (x, y) and $(0, \varepsilon)$ with the three-dimensional real space

$$\{(p, q) \in \mathbf{C}^2 : Re(q) = 0\},$$

where $Re(q)$ and $Im(q)$ are the real and imaginary parts of q. The point $(0, \varepsilon)$ is mapped to ∞. Explicitly

$$(x, y) \mapsto \begin{cases} \left(\frac{\varepsilon Re(x)}{\varepsilon - Re(y)}, \frac{\varepsilon Im(x)}{\varepsilon - Re(y)}, \frac{\varepsilon Im(y)}{\varepsilon - Re(y)}\right) & \text{if } Re(y) \neq \varepsilon \\ \infty & \text{if } Re(y) = \varepsilon \end{cases}$$

with inverse

$$\begin{aligned} (u, v, w) &\mapsto \varepsilon\left(\frac{2\varepsilon(u + iv)}{u^2 + v^2 + w^2 + \varepsilon^2}, \frac{u^2 + v^2 + w^2 - \varepsilon^2 + 2iw\varepsilon}{u^2 + v^2 + w^2 + \varepsilon^2}\right) \\ \infty &\mapsto (0, \varepsilon). \end{aligned}$$

Example 1.1 Let C be the complex algebraic curve defined by

$$xy = y^2$$

which is the union of the complex lines defined by $y = 0$ and $x = y$. Under stereographic projection, as above, the intersections of these complex lines with the sphere

$$S = \{(x, y) \in \mathbf{C}^2 : \mid x \mid^2 + \mid y \mid^2 = \varepsilon^2\}$$

are mapped to the circle

$$\{(u, v, w) \in \mathbf{R}^3 : w = 0, u^2 + v^2 = \varepsilon^2\}$$

and the ellipse

$$\{(u, v, w) \in \mathbf{R}^3 : v = w, (u - \varepsilon)^2 + 2v^2 = 2\varepsilon^2\}$$

in $\mathbf{R}^3 \cup \{\infty\}$. This circle and ellipse are *linked* in the sense that neither can be continuously shrunk to a point without passing through the other (see figure 1.5).

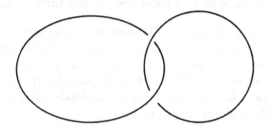

Figure 1.5: A link

Example 1.2 Now let C be the cubic curve defined by

$$y^2 = x^3.$$

Any point $(x, y) \in C$ can be written as

$$(x, y) = (s^2, s^3)$$

for a unique $s \in \mathbf{C}$, and then (x, y) belongs to the sphere

$$S = \{(x, y) \in \mathbf{C}^2 \; : |\, x\,|^2 + |\, y\,|^2 = \varepsilon^2\}$$

if and only if

$$|\, s\,| = \delta$$

where δ is the unique positive solution to the equation

$$\delta^4 + \delta^6 = \varepsilon^2.$$

Therefore

$$C \cap S = \{(\delta^2 e^{2it}, \delta^3 e^{3it}) \; : \; t \in [0, 2\pi)\}$$

which is contained in the subset

$$\{(x, y) \in \mathbf{C}^2 \; : |\, x\,| = \delta^2, |\, y\,| = \delta^3\}$$

of S. Under stereographic projection this subset is mapped onto the subset of $\mathbf{R}^3$ consisting of all those $(u, v, w) \in \mathbf{R}^3$ satisfying

$$2\varepsilon^2 \sqrt{u^2 + v^2} = \delta^2(u^2 + v^2 + w^2 + \varepsilon^2)$$

or equivalently

$$(\sqrt{u^2 + v^2} - \varepsilon^2 \delta^{-2})^2 + w^2 = \varepsilon^2 \delta^2.$$

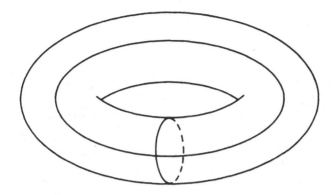

Figure 1.6: A torus

This is the surface T in $\mathbf{R}^3$ swept out by rotating the circle with equation

$$(v - \epsilon^2\delta^{-2})^2 + w^2 = \epsilon^2\delta^2 \tag{1.5}$$

about the w-axis. Topologically T is a torus— that is, the surface of a ring (see figure 1.6). The image of $C \cap S$ under stereographic projection is a knot in $\mathbf{R}^3$ which lies on the torus T. As we travel along the points

$$(\delta^2 e^{2it}, \delta^3 e^{3it})$$

of $C \cap S$ with the parameter t varying from 0 to 2π, the angle of rotation about the w-axis of the corresponding points in T is given by

$$\tan^{-1}\left(\frac{v}{u}\right) = \tan^{-1}\left(\frac{Im(x)}{Re(x)}\right) = \arg(x) = 2t \pmod{2\pi}$$

which varies from 0 to 2π twice. Thus the knot on T winds twice around the w-axis. On the other hand the standard angular coordinate on the circle defined by equation (1.5) rotated through some fixed angle about the w-axis is

$$\tan^{-1}\left(\frac{w}{\sqrt{u^2+v^2}-\epsilon^2\delta^{-2}}\right) = \tan^{-1}\left(\frac{Im(y)}{|x|-\epsilon\delta^{-2}(\epsilon-Re(y))}\right)$$
$$= \tan^{-1}\left(\frac{\sin(3t)}{\sqrt{1+\delta^2}\cos(3t)-\delta}\right),$$

and it is not difficult to see that when δ is small enough this varies from 0 to 2π three times as t varies from 0 to 2π. This means that the image of $C \cap S$ under stereographic projection is a *trefoil* knot on the torus T (see figure 1.7). Only two points on this knot can be seen in the real picture (figure 1.8).

Algebraic singularities are important in many parts of mathematics besides knot theory (e.g. catastrophe theory, which has been applied to areas as diverse as physics, ecology and economics. For a layman's introduction see V. Arnol'd's book [Arnol'd 86] or [Brieskorn & Knörrer 86] p.52).

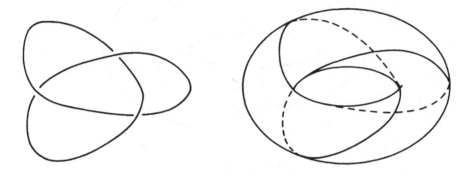

Figure 1.7: A trefoil knot in the plane and a trefoil knot on a torus

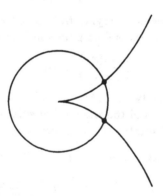

Figure 1.8: The real points on the curve $y^2 = x^3$ and a small sphere about the origin

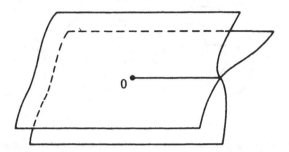

Figure 1.9: Two copies of the cut plane glued along $[0, \infty)$

1.2.3 Complex analysis

There is no holomorphic single-valued function $\sqrt{z}$ defined on the whole complex plane $\mathbf{C}$. However if we cut $\mathbf{C}$ along the non-negative real axis $[0, \infty)$ we can define two holomorphic functions $\pm\sqrt{z}$ on $\mathbf{C} - [0, \infty)$ by

$$\pm\sqrt{z} = \pm\sqrt{r}e^{i\frac{\theta}{2}} \quad \text{if} \quad z = re^{i\theta}, 0 < \theta < 2\pi, r > 0.$$

Here $\sqrt{r}$ means the positive square root of r. Note that if $r \in [0, \infty)$ then $+\sqrt{z}$ tends to $\sqrt{r}$ and $-\sqrt{z}$ tends to $-\sqrt{r}$ as z tends to r through values in the upper half plane, whereas $+\sqrt{z}$ tends to $-\sqrt{r}$ and $-\sqrt{z}$ tends to $\sqrt{r}$ as z tends to r through values in the lower half plane.

We can take two copies of $\mathbf{C}$ cut along $[0, \infty)$ and "glue" the upper side of the cut in the first copy to the lower side of the cut in the second copy, and the lower side of the cut in the first copy to the upper side of the cut in the second copy. The picture in figure 1.9 can help to give an intuitive idea of what the space X obtained in this way looks like. However this picture is misleading in that it gives the impression that the space intersects itself along the positive real axis, which is not supposed to happen. A better topological picture of X with an extra point added at infinity is given as follows. Think of two copies of the complex sphere $\mathbf{C} \cup \{\infty\}$ cut along the positive real axis from 0 to ∞. Open up the cuts and glue the two copies together to get $X \cup \{\infty\}$ which is again a sphere, topologically (see figure 1.10). On this space X it makes sense to say that there is a single-valued holomorphic function $\sqrt{z}$ defined by $+\sqrt{z}$ on the first copy of the cut plane and by $-\sqrt{z}$ on the second (see Chapter 5). If we add in a point ∞ and set $\sqrt{\infty} = \infty$ then we get a holomorphic function from $X \cup \{\infty\}$ to $\mathbf{C} \cup \{\infty\}$ which takes every value except 0 and ∞ exactly twice.

We can either construct this space $X \cup \{\infty\}$ abstractly, or we can think of it as being the complex algebraic curve

$$\{(z, w) \in \mathbf{C}^2 \ : \ w^2 = z\}$$

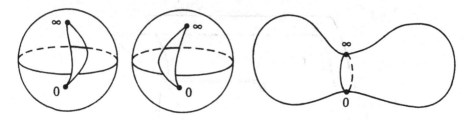

Figure 1.10: Two copies of the complex sphere glued along $[0, \infty)$

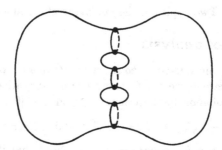

Figure 1.11: Two copies of the complex sphere glued along three cuts

where one copy of $\mathbf{C} - [0, \infty)$ corresponds to

$$\{(z, w) \in \mathbf{C}^2 \ : \ w = +\sqrt{z}\}$$

and the other to

$$\{(z, w) \in \mathbf{C}^2 \ : \ w = -\sqrt{z}\}.$$

Similarly to make a space on which the function

$$\sqrt{(z - \alpha_1)(z - \alpha_2) \dots (z - \alpha_k)}$$

is single-valued (where $\alpha_1, \dots \alpha_k$ are distinct complex numbers) one cuts the complex sphere from α_1 to α_2 , from α_3 to α_4 and so on (including ∞ if k is odd) and glues two copies of the result together as before, to get a space Y which is topologically a sphere with $[\frac{1}{2}(k - 1)]$ handles. (Figures 1.11 and 1.12 are both topological representations of the case $k = 6$). This space can also be thought of as the curve

$$\{(z, w) \in \mathbf{C}^2 \ : \ w^2 = (z - \alpha_1) \dots (z - \alpha_k)\}$$

together with one or two "points at infinity" (depending on whether k is odd or even).

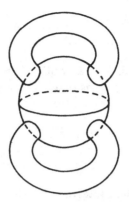

Figure 1.12: A sphere with two handles

We can play the same game for any multi-valued holomorphic function $w(z)$ satisfying a relation

$$P(z, w(z)) = 0$$

where $P(x, y)$ is an irreducible polynomial. The space we construct is just the complex algebraic curve

$$C = \{(z, w) \in \mathbf{C}^2 \; : \; P(z, w) = 0\}$$

minus the points where $\frac{\partial P}{\partial w}(z, w)$ vanishes (sometimes called "branch points" or "ramification points"). We can then add in points to fill the gaps left by the branch points and suitable "points at infinity" to get a compact space on which it makes sense to say that $w(z)$ is a single-valued holomorphic function with values in $\mathbf{C} \cup \{\infty\}$ (see §7.1).

1.2.4 Abelian integrals.

For our last example of how complex algebraic curves appear in other parts of mathematics we begin by considering the integral

$$\int_a^b r(x)\, dx$$

where $r(x)$ is a rational function of x, i.e.

$$r(x) = \frac{p(x)}{q(x)}$$

where $p(x)$ and $q(x)$ are polynomials in x and $q(x)$ is not identically zero. Assume that $q(x) \neq 0$ if $x \in [a, b]$ so the integral exists. We can expand $r(x)$

in partial fractions

$$r(x) = r_0(x) + \alpha_1(x - \beta_1)^{-k_1} + \ldots + \alpha_m(x - \beta_m)^{-k_m}$$

where $r_0(x)$ is a polynomial, k_i is a positive number and α_i, β_i are complex numbers for $1 \le i \le m$. This means that $r(x)$ has an indefinite integral which is the sum of a rational function and terms

$$\alpha_i \log(x - \beta_i)$$

for each i such that $k_i = 1$.

Next consider the standard integral

$$\int_a^b (1 + x^2)^{-\frac{1}{2}} \, dx.$$

One is taught how to work out this integral by substituting either

$$x = \tan\theta$$

or

$$x = \frac{2t}{(1 - t^2)}.$$

(These two substitutions are related by $t = \tan\frac{1}{2}\theta$). The latter substitution

$$x = \frac{2t}{(1 - t^2)}$$

turns any integral of the form

$$\int_a^b R(x, \sqrt{1 + x^2}) \, dx$$

where $R(x, y) = \frac{P(x,y)}{Q(x,y)}$ is a rational function of two variables, into the integral of a rational function of t.

Similarly we can deal with any integral of the form

$$\int_a^b R(x, w(x)) \, dx$$

where $R(x, y)$ is rational and $w(x)$ is a continuous function of x in $[a, b]$ such that $w(x)$ and x satisfy a polynomial relation of degree two:

$$aw(x)^2 + (bx + c)w(x) + dx^2 + ex + f = 0, \qquad (1.6)$$

where $a, b, c, d, e, f \in \mathbf{C}$. We may assume that a is nonzero. Then

$$(w(x) + (bx + c)/2a)^2 = \begin{cases} \alpha(x - \lambda)(x - \mu), & \alpha \neq 0 \quad \text{(Case 1)} \\ \alpha(x - \lambda), & \alpha \neq 0 \quad \text{(Case 2)} \\ \alpha & \quad \text{(Case 3)} \end{cases}$$

for some $\alpha, \lambda, \mu \in \mathbf{C}$. On substituting

$$x = \begin{cases} \frac{(\mu - \lambda)it}{(1 - t^2)} + \frac{1}{2}(\lambda + \mu) & \text{(Case 1, } \lambda \neq \mu) \\ \lambda + \frac{t^2}{\alpha} & \text{(Case 2)} \\ t & \text{otherwise} \end{cases}$$

the integral

$$\int_a^b R(x, w(x)) \, dx$$

becomes the integral of a rational function of t. However more care is necessary now because these substitutions take us into the complex plane. It is important to remember that the integral of a function along a path γ in $\mathbf{C}$ usually depends on the path γ as well as its endpoints a and b. For example

$$\log y = \int_1^y x^{-1} \, dx$$

is a multi-valued function of $y \in \mathbf{C} - \{0\}$.

Definition 1.3 *An abelian integral is an integral of the form*

$$\int_\gamma R(z, w(z)) \, dz$$

where $\gamma : [0, 1] \to \mathbf{C}$ is a continuous path in $\mathbf{C}$, $R(x, y)$ is a rational function of two variables and $w(z)$ is a continuous function of z defined on the image of the path γ such that $w(z)$ and z satisfy a polynomial relation

$$P(z, w(z)) = 0.$$

Integrals of this form were extensively studied in the last century, in particular by the Norwegian mathematician Niels Hendrik Abel, after whom they are named. They occur in many problems in applied mathematics.

Example 1.4 If $w(z)$ satisfies

$$w(z)^2 = (z - \alpha_1) \ldots (z - \alpha_k)$$

where $\alpha_1, \ldots, \alpha_k$ are distinct complex numbers then

$$\int_\gamma R(z, w(z)) \, dz$$

is called an elliptic integral if k is 3 or 4 and a hyperelliptic integral if k is at least 5.

Example 1.5 The arc-length of an ellipse

$$\frac{x^2}{a^2} + \frac{y^2}{b^2} = 1$$

between points with x-coordinates c and d is

$$\frac{1}{a} \int_c^d \frac{a^4 + (b^2 - a^2)x^2}{\sqrt{(a^2 - x^2)(a^4 + (b^2 - a^2)x^2)}}\, dx$$

which is an elliptic integral (hence the name).

Example 1.6 Elliptic integrals also occur in solving the differential equation for a simple pendulum:

$$\ddot{\theta} = -k\sin\theta.$$

When multiplied by $2\dot{\theta}$ this equation can be integrated to give

$$(\dot{\theta})^2 = 2k\cos\theta + c$$

and on substituting $x = \cos\theta$ we get

$$\int dt = \int [(1 - x^2)(2kx + c)]^{-\frac{1}{2}}\, dx$$

which is an elliptic integral.

We have seen that when the total degree of the polynomial relation $P(x,y)$ between z and $w(z)$ is at most two then by a suitable rational substitution the abelian integral

$$\int_\gamma R(z, w(z))\, dz$$

can be turned into the integral of a rational function and hence can be evaluated. However when the degree of $P(x,y)$ is strictly greater than two this is no longer the case. We shall discuss why this is so, and why the degrees 0, 1 and 2 are special, by studying properties of the complex algebraic curve

$$\{(z, w) \in \mathbf{C}^2 \; : \; P(z, w) = 0\}$$

defined by P (see §6.1).

1.3   Real Algebraic Curves

The main object of this book is to study *complex* algebraic curves. However this short section is included on real algebraic curves. This is partly because real algebraic curves are very important in the history of mathematics, and there are many important classical examples. It is also partly because it can sometimes be useful to look at real curves when studying complex ones. However it is vital to stress that it is often dangerous to do this. There are many important differences between the theories of real and complex curves.

1.3.1 Hilbert's Nullstellensatz

One of the most important differences arises when we consider the question: "When do two polynomials define the same curve?"

Example 1.7 The polynomials

$$P(x,y) = x^5 + x^3 y = x^3(x^2 + y)$$

and

$$Q(x,y) = x^6 + 2x^4 y + x^2 y^2 = x^2(x^2 + y)^2$$

define the same real and complex algebraic curves in the sense that

$$\{(x,y) \in \mathbf{C}^2 \ : \ P(x,y) = 0\} = \{(x,y) \in \mathbf{C}^2 \ : \ Q(x,y) = 0\}$$

and

$$\{(x,y) \in \mathbf{R}^2 \ : \ P(x,y) = 0\} = \{(x,y) \in \mathbf{R}^2 \ : \ Q(x,y) = 0\}.$$

Example 1.8 The polynomials

$$P(x,y) = x^5 + x^3 y = x^3(x^2 + y)$$

and

$$R(x,y) = x^3 + xy + x^3 y^2 + xy^3 = x(x^2 + y)(y^2 + 1)$$

define the same *real* algebraic curves (since $y^2 + 1 \neq 0$ for all $y \in \mathbf{R}$) but not the same *complex* algebraic curves.

For complex algebraic curves we have a simple answer to the question when two polynomials define the same curve.

Theorem 1.9 (Hilbert's Nullstellensatz) *If $P(x,y)$ and $Q(x,y)$ are polynomials with complex coefficients then*

$$\{(x,y) \in \mathbf{C}^2 \ : \ P(x,y) = 0\} = \{(x,y) \in \mathbf{C}^2 \ : \ Q(x,y) = 0\}$$

if and only if there exist positive integers n and m such that P divides Q^n and Q divides P^m; or equivalently if and only if P and Q have the same irreducible factors, possibly occuring with different multiplicities.

Proof. A proof can be found in, for example [Atiyah & Macdonald 69] p.85. □

The analagous result does *not* hold for real algebraic curves, as example 1.8 shows.

1.3.2   Techniques for drawing real algebraic curves

It is important to keep stressing the fact that it can be dangerous to think about real algebraic curves when investigating the corresponding complex algebraic curves. Nonetheless it can sometimes be useful to sketch the real curves, and so in this section we shall review very briefly some basic techniques.

Let

$$C = \{(x,y) \in \mathbf{R}^2 \; : \; P(x,y) = 0\}$$

be a real algebraic curve. We can get some preliminary information about C by working out where it meets the axes; that is the points $(x,0)$ and $(0,y)$ such that

$$P(x,0) = 0 = P(0,y).$$

We can work out that the tangent lines to C at these points, and any other points we happen to know lie on the curve. The tangent line to C at (a,b) is defined by

$$(x - a)\frac{\partial P}{\partial x}(a,b) + (y - b)\frac{\partial P}{\partial y}(a,b) = 0.$$

We can find the points (a,b) where the tangent line is parallel to an axis, i.e. where $\frac{\partial P}{\partial x}(a,b) = 0$ or $\frac{\partial P}{\partial y}(a,b) = 0$. We can also work out the singular points of C, i.e. the points (a,b) where

$$\frac{\partial P}{\partial x}(a,b) = 0 = \frac{\partial P}{\partial y}(a,b)$$

so that there is no well-defined tangent direction.

We can also investigate what the curve C looks like near a singular point (a,b) as follows. Let us assume for simplicity that (a,b) is the origin $(0,0)$; in general we can substitute $x - a$ for x and $y - b$ for y. Let m be the multiplicity of C at $(0,0)$; that is, the smallest value of $i + j$ such that $x^i y^j$ occurs with nonzero coefficient a_{ij} in the polynomial

$$P(x,y) = \sum_{i,j \geq 0} a_{ij} x^i y^j.$$

Of course, $(0,0)$ is a singular point if and only if $m > 1$. Near the origin terms of lowest order in x and y dominate so we can hope that C looks approximately like the curve defined by

$$\sum_{i+j=m} a_{ij} x^i y^j = 0.$$

Figure 1.13: The eight curve

Since the polynomial

$$\sum_{i+j=m} a_{ij}x^i y^j$$

is *homogeneous* of degree m in x and y it factorises over $\mathbf{C}$ as a product of m linear polynomials

$$\alpha_i x + \beta_i y$$

for $1 \le i \le m$ (see lemma 2.8). Let us suppose that the ratios $\alpha_i : \beta_i$ are real for $1 \le i \le k$ and not real for $k < i \le m$ where $0 \le k \le m$. Then near the origin it is reasonable to hope that C looks "to a first approximation" like the union of the lines defined by

$$\alpha_i x + \beta_i y = 0$$

These are called the tangent lines to C at the origin. For example the "eight curve" or "lemniscate of Gerono" defined by

$$x^4 = x^2 - y^2$$

has a singularity at the origin of multiplicity $m = 2$ with tangent lines defined by $x = y$ and $x = -y$ (see figure 1.13). However things are not always quite as simple as this. Consider the curve defined by

$$x^2 y^2 = x^4 + y^2$$

which has a singular point at the origin of multiplicity $m = 2$. The lowest order term in the equation is y^2, so the tangent lines at the origin are both the x-axis $y = 0$. But the origin is actually an isolated point of the curve, since

$$x^2 y^2 = x^4 + y^2 \Rightarrow x^2 y^2 \ge y^2 \Rightarrow y = 0 \quad \text{or} \quad x^2 \ge 1$$

and the only point of the curve satisfying $y = 0$ is the origin.

For another example consider the cubic curve defined by $y^2 = x^3$ (see figure 1.1). Again the curve has a singular point at the origin and the lowest order term is y^2. In this case

$$x^3 = y^2 \Rightarrow x \ge 0$$

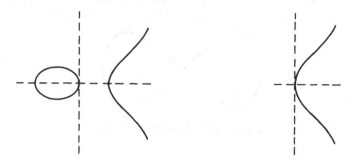

Figure 1.14: The curves $y^2 = x^3 + x$ and $y^2 = x^3 - x$

and near the origin the curve only looks like "half" the tangent lines which are both defined by $y = 0$.

These two examples show that care must be taken with the interpretation of the tangent lines to a real algebraic curve at a singular point.

1.3.3　Real algebraic curves inside complex algebraic curves

It is also important to note that rather different real algebraic curves can sit inside equivalent complex algebraic curves. For example the real cubic curve defined by

$$y^2 = x^3 + x$$

has two connected components whereas the real cubic curve defined by

$$y^2 = x^3 - x$$

has only one (see figure 1.14).However the corresponding complex curves are equivalent under the complex change of coordinates

$$(x, y) \mapsto (ix, e^{\frac{3\pi i}{4}} y).$$

1.3.4　Important examples of real algebraic curves

Now let us consider some examples of real algebraic curves which were important in antiquity.

Let

$$C = \{(x, y) \in \mathbf{R}^2 \ : \ P(x, y) = 0\}$$

be any real algebraic curve in $\mathbf{R}^2$, let q be a fixed point in $\mathbf{R}^2$ and let $a > 0$ be a fixed constant. Then the locus of all points $p \in \mathbf{R}^2$ such that the line in $\mathbf{R}^2$

through p and q meets C at a point of distance a from p is called a *conchoid* of C with respect to q and with parameter a. There are many classical examples of conchoids.

The *conchoid of Nicomedes* (c. 225 B.C.) is the conchoid of a line with respect to a point not on the line. If the line is defined by

$$x = b$$

and the point is the origin $(0,0)$ then the conchoid has equation

$$(x^2 + y^2)(x - b)^2 = a^2 x^2.$$

One can see from their definition that conchoids of Nicomedes can be drawn with the aid of a simple apparatus. This apparatus was used by the Greeks to trisect angles (this was a famous problem: cf. §1.1). They showed that it is possible to trisect an arbitrary angle α, given as the angle between two lines meeting at a point q, as follows. First one draws the perpendicular L to one line from a point r on the other line. Then one draws the conchoid of L with respect to q and with parameter

$$a = 2l$$

where l is the distance of r from q. The parallel to the first line through r meets the conchoid on the side away from q at a point p, and the line through p and q trisects the angle α (see figure 1.15).

The *limacon of Pascal* (named after the seventeenth century French mathematician Pascal) is the conchoid of a circle of radius b with respect to a point q on its circumference which we may take to be the origin. If we take the centre of the circle to lie on the x-axis, then the equation is

$$(x^2 + y^2 - 2bx)^2 = a^2(x^2 + y^2).$$

The limacon is the path traced by a point rigidly attached to a circle rolling on an equal fixed circle. More generally if the circles are allowed to have different radii b and c one obtains a curve called a *epitrochoid* or *hypotrochoid*, depending on whether the rolling circle is outside or inside the fixed circle. When the attached point actually lies on the rolling circle one has an *epicycloid* or *hypocycloid*. Epicycloids and hypocycloids were used (probably as early as the seventeenth century) to design the so-called cycloidal gear.

Important special cases of epicycloids and hypocycloids are the cardioid (an epicycloid with $b = c$), so called because of its heartlike shape, the

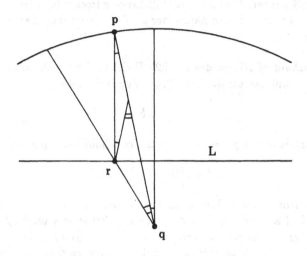

Figure 1.15: Trisecting an angle

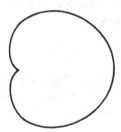

Figure 1.16: Cardioid

nephroid or kidney curve (an epicycloid with $c = 2b$), the deltoid (a hypocy-cloid with $3b = c$) and the astroid (a hypocycloid with $4b = c$). Their equations in suitable coordinate systems are

$$
\begin{aligned}
(x^2 + y^2 - 2bx)^2 &= 4b^2(x^2 + y^2) && \text{Cardioid} \\
(x^2 + y^2 - 4b^2)^3 &= 108b^4 y^2 && \text{Nephroid} \\
(x^2 + y^2)^2 - 8bx(x^2 - 3y^2) + 18b^2(x^2 + y^2) &= 27b^4 && \text{Deltoid} \\
(x^2 + y^2 - b^2)^3 + 27b^2 x^2 y^2 &= 0. && \text{Astroid}
\end{aligned}
$$

Examples are sketched in figures 1.16–1.19.   The degenerate case of an epicycloid which occurs when the fixed circle is replaced by a line is called a *cycloid* (see figure 1.20). It was discovered by Bernoulli and others (around 1700) that given two points in a vertical plane, the curve along which a particle takes the least time to slide from one point to the other under gravity is a cycloid. This discovery led to the development of the theory of variations.

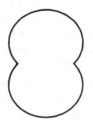

Figure 1.17: Nephroid

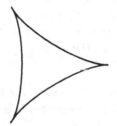

Figure 1.18: Deltoid

Figure 1.19: Astroid

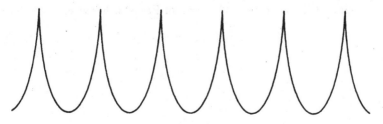

Figure 1.20: Cycloid

Another curve discovered by the Greeks is the *cissoid of Diocles*. A cissoid is associated to the choice of two curves C and D and a fixed point q. The cissoid of C and D with respect to q is the locus of all points p such that the line through p and q meets C and D in two points whose distance apart is the same as the distance between p and q. The cissoid of Diocles is the cissoid of a circle and a tangent line with respect to the point on the circumference of the circle opposite the point of tangency. If we take this point to be the origin and the tangent line to be defined by $x = a$ then the equation of the cissoid is

$$y^2(a - x) = x^3.$$

This cissoid can be drawn using, for example, an apparatus due to Newton. A right angle with two arms of length a is moved in the plane in such a way that the endpoint of one arm moves on the line defined by $x = \frac{1}{2}a$ and the other arm always passes through the point $(-\frac{1}{2}a, 0)$. Then the midpoint of the first arm describes the cissoid.

Diocles was able to use his cissoid to solve the Delian problem of doubling the cube (that is, constructing a line segment of length $2^{\frac{1}{3}}$ times a given unit length). The intersection of the cissoid

$$y^2(1 - x) = x^3$$

with the line defined by

$$x + 2y = 1$$

is the point $(2^{\frac{1}{3}}\gamma, \gamma)$ where $\gamma = (2 + 2^{\frac{1}{3}})^{-1}$. Using this and standard ruler and compass constructions a line segment of length $2^{\frac{1}{3}}$ units can be drawn.

Real algebraic curves important in mechanics are the *caustic curves* investigated by Tschirnhausen and Huygens at the end of the seventeenth century. A catacaustic (respectively diacaustic) of a given curve C is a curve which is tangent to all the light rays from a given source after reflection (respectively refraction) in C. If the light source is at infinity the incoming rays are parallel. For example the catacaustic of a circle C with source q is a cardioid if $q \in C$, a nephroid if $q = \infty$ and a limacon of Pascal otherwise.

Chapter 2

Foundations

This chapter contains the basic definitions and material we shall need to study complex algebraic curves. We shall first define complex algebraic curves in $\mathbf{C}^2$ and then add "points at infinity" to obtain complex projective curves.

2.1 Complex algebraic curves in $\mathbf{C}^2$

Let $P(x, y)$ be a non-constant polynomial in two variables with complex co-efficients. We say that $P(x, y)$ has no repeated factors if we cannot write

$$P(x, y) = (Q(x, y))^2 R(x, y)$$

where $Q(x, y)$ and $R(x, y)$ are polynomials and $Q(x, y)$ is nonconstant.

Definition 2.1 *Let $P(x, y)$ be a nonconstant polynomial in two variables with complex coefficients and no repeated factors. Then the complex algebraic curve in $\mathbf{C}^2$ defined by $P(x, y)$ is*

$$C = \{(x, y) \in \mathbf{C}^2 \ : \ P(x, y) = 0\}.$$

The reason for the assumption in this definition that $P(x, y)$ has no repeated factors is the theorem called Hilbert's Nullstellensatz already quoted in §1.3.

Theorem (Hilbert's Nullstellensatz) *If $P(x, y)$ and $Q(x, y)$ are polynomials with complex coefficients then*

$$\{(x, y) \in \mathbf{C}^2 \ : \ P(x, y) = 0\} = \{(x, y) \in \mathbf{C}^2 \ : \ Q(x, y) = 0\}$$

if and only if there exist positive integers m and n such that P divides Q^n and Q divides P^m; or equivalently if and only if P and Q have the same irreducible factors, possibly occuring with different multiplicities.

Corollary 2.2 *If $P(x,y)$ and $Q(x,y)$ have no repeated factors then they define the same complex algebraic curve in $\mathbf{C}^2$ if and only if they are scalar multiples of each other, i.e.*

$$P(x,y) = \lambda Q(x,y)$$

for some $\lambda \in \mathbf{C} - \{0\}$.

Proof. This is an immediate consequence of Hilbert's Nullstellensatz. $\square$

Remark 2.3 A more general way of defining a complex algebraic curve in $\mathbf{C}^2$ is as an equivalence class of nonconstant polynomials in two variables, where two polynomials are equivalent if and only if they are scalar multiples of each other. A polynomial with repeated factors is then thought of as defining a curve with multiplicities attached; for example $(y - x^2)^3$ defines the same curve as $y - x^2$ but with multiplicity three. Through most of this book the definition 2.1 will be sufficient; occasionally however (e.g. in §3.1) we shall need to allow the use of curves defined by polynomials with repeated factors.

Definition 2.4 *The* degree *d of the curve C defined by $P(x,y)$ is the degree of the polynomial P, i.e.*

$$d = \max \{r + s \ : \ c_{r,s} \neq 0\}$$

where

$$P(x,y) = \sum_{r,s} c_{r,s} x^r y^s.$$

A point $(a,b) \in C$ is called a singular *point (or singularity) of C if*

$$\frac{\partial P}{\partial x}(a,b) = 0 = \frac{\partial P}{\partial y}(a,b).$$

The set of singular points of C is denoted Sing(C). C is called nonsingular *if $Sing(C) = \phi$.*

Examples 2.5 The curve defined by $x^2 + y^2 = 1$ is nonsingular. The curve defined by $y^2 = x^3$ has one singular point $(0,0)$.

Definition 2.6 *A curve defined by a linear equation*

$$\alpha x + \beta y + \gamma = 0,$$

where α, β, γ are complex numbers and α and β are not both zero, is called a line.

At this point we must recall an important definition and lemma.

2.1. COMPLEX ALGEBRAIC CURVES IN $\mathbf{C}^2$

Definition 2.7 *A nonzero polynomial $P(x_1, \ldots, x_n)$ in n variables is homogeneous of degree d if*

$$P(\lambda x_1, \ldots, \lambda x_n) = \lambda^d P(x_1, \ldots, x_n)$$

for all $\lambda \in \mathbf{C}$. Equivalently P has the form

$$P(x_1, \ldots, x_n) = \sum_{r_1 + \ldots + r_n = d} a_{r_1 \ldots r_n} x_1^{r_1} \ldots x_n^{r_n}$$

for some complex numbers $a_{r_1 \ldots r_n}$.

Note that every polynomial factor $Q(x_1, \ldots, x_n)$ of a homogeneous polynomial $P(x_1, \ldots, x_n)$ is also homogeneous (see Appendix A).

Lemma 2.8 *If $P(x, y)$ is a nonzero homogeneous polynomial of degree d in two variables with complex coefficients then it factors as a product of linear polynomials*

$$P(x, y) = \prod_{i=1}^{d} (\alpha_i x + \beta_i y)$$

for some $\alpha_i, \beta_i \in \mathbf{C}$.

Proof. We can write

$$P(x, y) = \sum_{r=0}^{d} a_r x^r y^{d-r} = y^d \sum_{r=0}^{d} a_r \left(\frac{x}{y}\right)^r$$

where $a_0, \ldots, a_d \in \mathbf{C}$ are not all zero. Let e be the largest element of $\{0, \ldots, d\}$ such that $a_e \neq 0$. Then

$$\sum_{r=0}^{d} a_r \left(\frac{x}{y}\right)^r$$

is a polynomial with complex coefficients of degree e in one variable $\frac{x}{y}$ and so can be factorised as

$$\sum_{r=0}^{d} a_r \left(\frac{x}{y}\right)^r = a_e \prod_{i=1}^{e} \left(\frac{x}{y} - \gamma_i\right)$$

for some $\gamma_1, \ldots, \gamma_e \in \mathbf{C}$. Then

$$\begin{aligned} P(x, y) &= a_e y^d \prod_{i=1}^{e} \left(\frac{x}{y} - \gamma_i\right) \\ &= a_e y^{d-e} \prod_{i=1}^{e} (x - \gamma_i y). \end{aligned}$$

The result follows. $\square$

Note that since $P(x, y)$ is a polynomial it has a finite Taylor expansion

$$P(x, y) = \sum_{i, j \geq 0} \frac{\partial^{i+j} P}{\partial x^i \partial y^j}(a, b) \frac{(x-a)^i (y-b)^j}{i! j!}$$

about any point (a, b) (see exercise 2.3).

Definition 2.9 *The* multiplicity *of the curve C defined by $P(x,y)$ at a point $(a,b) \in C$ is the smallest positive integer m such that*

$$\frac{\partial^m P}{\partial x^i \partial y^j}(a,b) \neq 0$$

for some $i \geq 0, j \geq 0$ such that $i + j = m$. The polynomial

$$\sum_{i+j=m} \frac{\partial^m P}{\partial x^i \partial y^j}(a,b) \frac{(x-a)^i (y-b)^j}{i! j!} \qquad (2.1)$$

is then homogeneous of degree m and so by lemma 2.8 can then be factored as the product of m linear polynomials of the form

$$\alpha(x-a) + \beta(y-b)$$

where $(\alpha, \beta) \in \mathbf{C}^2 - \{(0,0)\}$. The lines defined by these linear polynomials are called the tangent lines *to C at (a,b). The point (a,b) is nonsingular if and only if its multiplicity m is 1; in this case C has just one tangent line at (a,b) defined by*

$$\frac{\partial P}{\partial x}(a,b)\,(x-a) + \frac{\partial P}{\partial y}(a,b)\,(y-b) = 0.$$

A point $(a,b) \in C$ is called a double point *(respectively* triple point*, etc.) if its multiplicity is two (respectively three, etc.). A singular point (a,b) is called* ordinary *if the polynomial (2.1) has no repeated factors, i.e. if C has m distinct tangent lines at (a,b).*

Example 2.10 The cubic curves defined by

$$y^2 = x^3 + x^2$$

and

$$y^2 = x^3$$

have double points at the origin; the first is an ordinary double point but the second is not (see figure 2.1). The curve defined by

$$(x^4 + y^4)^2 = x^2 y^2$$

has a singular point of multiplicity four at the origin which is not ordinary; the curve defined by

$$(x^4 + y^4 - x^2 - y^2)^2 = 9x^2 y^2$$

has an ordinary singular point of multiplicity four (see figure 2.2).

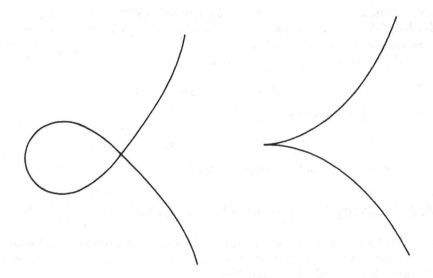

Figure 2.1: The curves $y^2 = x^3 + x^2$ and $y^2 = x^3$

Figure 2.2: The curves $(x^4 + y^4)^2 = x^2 y^2$ and $(x^4 + y^4 - x^2 - y^2)^2 = 9x^2 y^2$

Definition 2.11 *A curve* C *defined by a polynomial* $P(x,y)$ *is called* irreducible *if the polynomial is irreducible; that is, if* $P(x,y)$ *has no factors other than constants and scalar multiples of itself.*

If the irreducible factors of $P(x,y)$ *are*

$$P_1(x,y), \ldots, P_k(x,y)$$

then the curves defined by

$$P_1(x,y), \ldots, P_k(x,y)$$

are called the (irreducible) components *of* C.

2.2 Complex projective spaces

A complex algebraic curve C in $\mathbf{C}^2$ is never compact (see exercise 2.7). Recall some of the important properties of the topological condition of compactness (cf. e.g. Chapter 5 of [Sutherland 75]).

Properties 2.12
(i) A subset of $\mathbf{R}^n$ *or* $\mathbf{C}^n$ *is compact if and only if it is closed and bounded (Heine-Borel theorem).*
(ii) If $f : X \to Y$ *is a continuous map between topological spaces and* X *is compact, then* $f(X)$ *is compact.*
(iii) It follows from (i) and (ii) that if X *is a compact topological space and* $f : X \to \mathbf{R}$ *is a continuous function then* f *is bounded and attains its bounds.*
(iv) A closed subset of a compact space is compact.
(v) A compact subset of a Hausdorff space is closed.
(vi) A finite union of compact spaces is compact.

For many purposes it is useful to compactify complex algebraic curves in $\mathbf{C}^2$ by adding "points at infinity". For example, suppose we wish to study the intersection points of two curves, such as

$$y^2 = x^2 - 1, \quad y = cx$$

where c is a complex number. If $c \neq \pm 1$ these curves meet in two points. When $c = \pm 1$ they do not intersect, but they are asymptotic as x and y tend to infinity (cf. figure 2.3). We want to add points at infinity to $\mathbf{C}^2$ in such a way that the curves $y^2 = x^2 + 1$ and $y = cx$ intersect "at infinity" when $c = \pm 1$. Similarly parallel lines should "meet at infinity". More generally we would like any curves which are asymptotic to "meet at infinity".

To make this precise we use the concept of a projective space. The idea is to identify each $(x,y) \in \mathbf{C}^2$ with the one-dimensional complex linear subspace

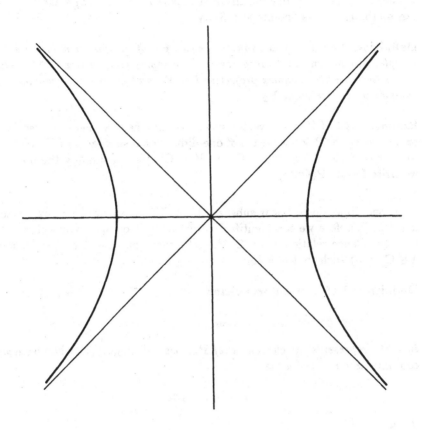

Figure 2.3: The curve $y^2 = x^2 - 1$ and the lines $y = \pm x$

of $\mathbf{C}^3$ spanned by $(x, y, 1)$. Every one-dimensional linear subspace of $\mathbf{C}^3$ which does not lie in the plane $\{(x, y, z) \in \mathbf{C}^3 : z = 0\}$ contains a unique point of the form $(x, y, 1)$. The one-dimensional subspaces of $\{(x, y, z) \in \mathbf{C}^3 : z = 0\}$ can be thought of as "points at infinity".

Definition 2.13 *Complex projective space* $\mathbf{P}_n$ *of dimension n is the set of complex one-dimensional subspaces of the complex vector space* $\mathbf{C}^{n+1}$. *When* $n = 1$ *we have the complex projective line* $\mathbf{P}_1$ *and when* $n = 2$ *we have the complex projective plane* $\mathbf{P}_2$.

Remark 2.14 If V is any vector space over any field K then the associated projective space $\mathbf{P}(V)$ is the set of one-dimensional subspaces of V. We shall work only with the case $K = \mathbf{C}$ and $V = \mathbf{C}^{n+1}$ so to simplify the notation we write $\mathbf{P}_n$ for $\mathbf{P}(\mathbf{C}^{n+1})$.

A one-dimensional linear subspace U of $\mathbf{C}^{n+1}$ is spanned by any nonzero $u \in U$. Therefore we can identify $\mathbf{P}_n$ with the set of equivalence classes for the equivalence relation $\sim$ on $\mathbf{C}^{n+1} - \{0\}$ such that $a \sim b$ if there is some $\lambda \in \mathbf{C} - \{0\}$ such that $a = \lambda b$.

Definition 2.15 *Any nonzero vector*

$$(x_0, \ldots, x_n)$$

in $\mathbf{C}^{n+1}$ *represents an element* x *of* $\mathbf{P}_n$*: we call* $(x_0, \ldots, x_n)$ *homogeneous* coordinates *for* x *and write*

$$x = [x_0, \ldots, x_n].$$

Then

$$\mathbf{P}_n = \{[x_0, \ldots, x_n] : (x_0, \ldots, x_n) \in \mathbf{C}^{n+1} - \{0\}\}$$

and

$$[x_0, \ldots, x_n] = [y_0, \ldots, y_n]$$

if and only if there is some $\lambda \in \mathbf{C} - \{0\}$ *such that* $x_j = \lambda y_j$ *for all* j.

Now we can make $\mathbf{P}_n$ into a topological space (we shall show that unlike $\mathbf{C}^n$ it is compact). We define $\Pi : \mathbf{C}^{n+1} - \{0\} \to \mathbf{P}_n$ by

$$\Pi(x_0, \ldots, x_n) = [x_0, \ldots, x_n]$$

and give $\mathbf{P}_n$ the *quotient topology* induced from the usual topology on $\mathbf{C}^{n+1} - \{0\}$ (cf. [Sutherland 75] p.68). That is, a subset A of $\mathbf{P}_n$ is open if and only if $\Pi^{-1}(A)$ is an open subset of $\mathbf{C}^{n+1} - \{0\}$.

Remark 2.16 Note that

(i) a subset B of $\mathbf{P}_n$ is closed if and only if $\Pi^{-1}(B)$ is a closed subset of $\mathbf{C}^{n+1} - \{0\}$;

(ii) $\Pi : \mathbf{C}^{n+1} - \{0\} \to \mathbf{P}_n$ is continuous;

(iii) if X is any topological space a function $f : \mathbf{P}_n \to X$ is continuous if and only if

$$f \circ \Pi : \mathbf{C}^{n+1} - \{0\} \to X$$

is continuous; more generally if A is any subset of $\mathbf{P}_n$ then a function $f : A \to X$ is continuous if and only if

$$f \circ \Pi : \Pi^{-1}(A) \to X$$

is continuous.

We define subsets $U_0, \ldots, U_n$ of $\mathbf{P}_n$ by

$$U_j = \{[x_0, \ldots, x_n] \in \mathbf{P}_n \; : \; x_j \neq 0\}.$$

Notice that the condition $x_j \neq 0$ is independent of the choice of homogeneous coordinates, and that

$$\Pi^{-1}(U_j) = \{(x_0, \ldots, x_n) \in \mathbf{C}^{n+1} \; : \; x_j \neq 0\}$$

is an open subset of $\mathbf{C}^{n+1} - \{0\}$, so U_j is an open subset of $\mathbf{P}_n$.

Define $\phi_0 : U_0 \to \mathbf{C}_n$ by

$$\phi_0[x_0, \ldots, x_n] = \left(\frac{x_1}{x_0}, \ldots, \frac{x_n}{x_0}\right).$$

This is a well-defined map with inverse

$$(y_1, \ldots, y_n) \mapsto [1, y_1, \ldots, y_n]$$

by definition 2.15. The coordinates $(y_1, \ldots, y_n)$ are called *inhomogeneous coordinates* on U_0.

It is easy to check using remark 2.16 that $\phi_0 : U_0 \to \mathbf{C}^n$ is continuous (we just have to show that its composition with $\Pi : \Pi^{-1}(U_0) \to U_0$ is continuous) and its inverse is the composition of Π with the continuous map from $\mathbf{C}^n$ to $\mathbf{C}^{n+1} - \{0\}$ given by

$$(y_1, \ldots, y_n) \mapsto (1, y_1, \ldots, y_n).$$

Thus ϕ_0 is a homeomorphism.

Similarly there are homeomorphisms $\phi_j : U_j \to \mathbf{C}^n$ for each j between 1 and n given by

$$\phi_j[x_0, \ldots, x_n] = \left(\frac{x_0}{x_j}, \ldots, \frac{x_{j-1}}{x_j}, \frac{x_{j+1}}{x_j}, \ldots, \frac{x_n}{x_j}\right).$$

Remark 2.17 The complement of U_n in $\mathbf{P}_n$ is the hyperplane

$$\{[x_0, \ldots, x_n] \in \mathbf{P}_n \; : \; x_n = 0\}.$$

This can be identified with $\mathbf{P}_{n-1}$ in the obvious way. Thus we can build up the projective spaces $\mathbf{P}_n$ inductively. $\mathbf{P}_0$ is a single point. $\mathbf{P}_1$ can be thought of as $\mathbf{C}$ together with a single point ∞ (i.e. a copy of $\mathbf{P}_0$) and thus can be identified with the Riemann sphere

$$\mathbf{C} \cup \{\infty\}.$$

$\mathbf{P}_2$ is $\mathbf{C}^2$ together with a "line at infinity" (i.e. a copy of $\mathbf{P}_1$), and in general $\mathbf{P}_n$ is $\mathbf{C}^n$ with a copy of $\mathbf{P}_{n-1}$ at infinity.

Since $\{U_j \; : \; 0 \leq j \leq n\}$ is an open cover of $\mathbf{P}_n$ and

$$\phi_j : U_j \to \mathbf{C}^n$$

is a homeomorphism for each j, a function f from $\mathbf{P}_n$ to a topological space X is continuous if and only if

$$f \circ \phi_j^{-1} : \mathbf{C}^n \to X$$

is continuous for each j. Similarly a function $f : X \to \mathbf{P}_n$ is continuous if and only if $f^{-1}(U_j)$ is open in X and

$$\phi_j \circ f : f^{-1}(U_j) \to \mathbf{C}^n$$

is continuous for each j.

Proposition 2.18 $\mathbf{P}_n$ *is compact.*

Proof. Let

$$S^{2n+1} = \{(x_0, \ldots, x_n) \in \mathbf{C}^{n+1} \; : \; |x_0|^2 + \ldots |x_n|^2 = 1\}.$$

Then S^{2n+1} is a sphere of dimension $2n + 1$. In particular it is a closed, bounded subset of $\mathbf{C}^{n+1}$, so by the Heine-Borel theorem 2.12(i) it is compact. The restriction $\Pi : S^{2n+1} \to \mathbf{P}_n$ of the map Π defined above is continuous, so its image is the image of a compact space under a continuous mapping, and hence is compact by 2.12(ii).

Now if $[x_0, \ldots, x_n] \in \mathbf{P}_n$ then

$$\lambda = |x_0|^2 + \ldots + |x_n|^2 > 0$$

so

$$[x_0, \ldots, x_n] = [\lambda^{-\frac{1}{2}} x_0, \ldots, \lambda^{-\frac{1}{2}} x_n].$$

But

$$| \lambda^{-\frac{1}{2}} x_0 |^2 + \ldots + | \lambda^{-\frac{1}{2}} x_n |^2 = 1$$

so

$$[x_0, \ldots, x_n] \in \Pi(S^{2n+1}).$$

Thus $\Pi : S^{2n+1} \to \mathbf{P}_n$ is surjective, and the result follows. □

Definition 2.19 *A* projective transformation *of* $\mathbf{P}_n$ *is a bijection*

$$f : \mathbf{P}_n \to \mathbf{P}_n$$

such that for some linear isomorphism $\alpha : \mathbf{C}^{n+1} \to \mathbf{C}^{n+1}$ *we have*

$$f[x_0, \ldots, x_n] = [y_0, \ldots, y_n]$$

where

$$(y_0, \ldots, y_n) = \alpha(x_0, \ldots, x_n),$$

i.e.

$$f \circ \Pi = \Pi \circ \alpha$$

where $\Pi : \mathbf{C}^{n+1} - \{0\} \to \mathbf{P}_n$ *is defined as before by*

$$\Pi(x_0, \ldots, x_n) = [x_0, \ldots, x_n].$$

Lemma 2.20 *A projective transformation* $f : \mathbf{P}_n \to \mathbf{P}_n$ *is continuous.*

Proof. We have

$$f \circ \Pi = \Pi \circ \alpha$$

for some linear isomorphism $\alpha : \mathbf{C}^{n+1} \to \mathbf{C}^{n+1}$. Since Π is continuous (by 2.16(ii)) and α is continuous it follows that $f \circ \Pi$ is continuous, and therefore by 2.16(iii) that f is continuous. □

Definition 2.21 *A* hyperplane *in* $\mathbf{P}_n$ *is the image under* $\Pi : \mathbf{C}^{n+1} - \{0\} \to \mathbf{P}_n$ *of* $V - \{0\}$ *where* V *is a subspace of* $\mathbf{C}^{n+1}$ *of dimension* n.

Lemma 2.22 *Given* $n + 2$ *distinct points* $p_0, \ldots, p_n$ *and* q *of* $\mathbf{P}_n$, *no* $n+1$ *of which lie on a hyperplane, there is a unique projective transformation taking* p_i *to* $[0, \ldots, 0, 1, 0, \ldots, 0]$ *where 1 is in the ith place, and taking* q *to* $[1, \ldots, 1]$.

Proof. Let $u_0, \ldots, u_n$ and v be elements of $\mathbf{C}^{n+1} - \{0\}$ whose images under Π are $p_0, \ldots, p_n$ and q. Then $u_0, \ldots, u_n$ form a basis of $\mathbf{C}^{n+1}$ so there is a unique linear transformation α of $\mathbf{C}^{n+1}$ taking $u_0, \ldots, u_n$ to the standard basis $(1, 0, \ldots, 0), \ldots, (0, \ldots, 0, 1)$. Moreover the condition on $p_0, \ldots, p_n$ and q implies that

$$\alpha(v) = (\lambda_1, \ldots, \lambda_n)$$

where $\lambda_1, \ldots, \lambda_n$ are all *nonzero* complex numbers. Thus the composition of α with the linear transformation defined by the diagonal matrix

$$\begin{pmatrix} \frac{1}{\lambda_1} & 0 & \cdots & 0 \\ 0 & \ddots & & 0 \\ \vdots & & \ddots & \vdots \\ 0 & \cdots & 0 & \frac{1}{\lambda_n} \end{pmatrix}$$

defines a projective transformation taking p_i to

$$[0, \ldots, 0, \frac{1}{\lambda_i}, 0, \ldots, 0] = [0, \ldots, 0, 1, 0, \ldots, 0]$$

and q to $[1,1,\ldots,1]$. Uniqueness is easy to check. □

Proposition 2.23 *The projective space* $\mathbf{P}_n$ *is Hausdorff.*

Proof. We need to show that if p and q are distinct points of $\mathbf{P}_n$ then they have disjoint open neighbourhoods in $\mathbf{P}_n$.

We have shown that there is a homeomorphism $\phi_0 : U_0 \to \mathbf{C}^n$ where U_0 is an open set of $\mathbf{P}_n$. If p and q lie in U_0 then $\phi_0(p)$ and $\phi_0(q)$ have disjoint open neighbourhoods V and W in $\mathbf{C}^n$ (since $\mathbf{C}^n$ is Hausdorff) and $\phi_0^{-1}(V)$ and $\phi_0^{-1}(W)$ are disjoint open neighbourhoods of p and q in $\mathbf{P}_n$. In particular this applies when $p = [1,0,\ldots, 0]$ and $q = [1,1,\ldots,1]$.

In general we can find points $p_0, \ldots, p_n$ of $\mathbf{P}_n$ such that $p_0 = p$ and no $n+1$ of the $n+2$ points $p_0, \ldots, p_n$ and q lie in a hyperplane. Then by lemma 2.22 there is a projective transformation $f : \mathbf{P}_n \to \mathbf{P}_n$ taking p to $[1,0,\ldots,0]$ and q to $[1,1,\ldots,1]$. We have already shown that $[1,0,\ldots,0]$ and $[1,1,\ldots,1]$ have disjoint open neighbourhoods $\phi_0^{-1}(V)$ and $\phi_0^{-1}(W)$ in $\mathbf{P}_n$. Since f is continuous (by lemma 2.20) and a bijection, the subsets $f^{-1}(\phi_0^{-1}(V))$ and $f^{-1}(\phi_0^{-1}(W))$ are disjoint open neighbourhoods of p and q in $\mathbf{P}_n$, as required. □

2.3   Complex projective curves in $\mathbf{P}_2$

Recall from §2.2 that the complex projective plane $\mathbf{P}_2$ is the set of one-dimensional complex subspaces of $\mathbf{C}^3$. We denote by $[x, y, z]$ the subspace spanned by $(x, y, z) \in \mathbf{C}^3 - \{0\}$, and thus

$$\mathbf{P}_2 = \{[x, y, z] \: : \: (x, y, z) \in \mathbf{C}^3 - \{0\} \}$$

and

$$[x, y, z] = [u, v, w]$$

if and only if there exists some $\lambda \in \mathbf{C} - \{0\}$ such that

$$x = \lambda u, \quad y = \lambda v, \quad z = \lambda w.$$

Recall also that a polynomial $P(x, y, z)$ is called homogeneous of degree d if

$$P(\lambda x, \lambda y, \lambda z) = \lambda^d P(x, y, z)$$

for all $\lambda \in \mathbf{C}$. Note that the first partial derivatives of P are homogeneous polynomials of degree $d - 1$.

Definition 2.24 *Let $P(x, y, z)$ be a nonconstant homogeneous polynomial in three variables x, y, z with complex coefficients. Assume that $P(x, y, z)$ has no repeated factors. Then the projective curve C defined by $P(x, y, z)$ is*

$$C = \{[x, y, z] \in \mathbf{P}_2 \ : \ P(x, y, z) = 0\}.$$

Note that the condition $P(x, y, z) = 0$ is independent of the choice of homogeneous coordinates (x, y, z) because P is a homogeneous polynomial and hence

$$P(\lambda x, \lambda y, \lambda z) = 0 \Leftrightarrow P(x, y, z) = 0$$

when $\lambda \in \mathbf{C} - \{0\}$.

Remark 2.25 Just as for curves in $\mathbf{C}^2$ it is in fact the case that two homogeneous polynomials $P(x, y, z)$ and $Q(x, y, z)$ with no repeated factors define the same projective curves in $\mathbf{P}_2$ if and only if they are scalar multiples of each other, and a homogeneous polynomial with repeated factors can be thought of as defining a curve with multiplicities attached to its components.

Definition 2.26 *The degree of a projective curve C in $\mathbf{P}_2$ defined by a homogeneous polynomial $P(x, y, z)$ is the degree d of $P(x, y, z)$. The curve C is called irreducible if $P(x, y, z)$ is irreducible, i.e. $P(x, y, z)$ has no nonconstant polynomial factors other than scalar multiples of itself. An irreducible projective curve D defined by a homogeneous polynomial $Q(x, y, z)$ is called a component of C if $Q(x, y, z)$ divides $P(x, y, z)$.*

Definition 2.27 *A point $[a, b, c]$ of a projective curve C in $\mathbf{P}_2$ defined by a homogeneous polynomial $P(x, y, z)$ is called singular if*

$$\frac{\partial P}{\partial x}(a, b, c) = \frac{\partial P}{\partial y}(a, b, c) = \frac{\partial P}{\partial z}(a, b, c) = 0.$$

The set of singular points of C is denoted by $Sing(C)$. The curve C is called nonsingular if $Sing(C) = \emptyset$.

Examples 2.28 The projective curve in $\mathbf{P}_2$ defined by $x^2 + y^2 = z^2$ is nonsingular.

The curve defined by $y^2z = x^3$ has one singular point [0,0,1].

Definition 2.29 *A projective curve defined by a linear equation*

$$\alpha x + \beta y + \gamma z = 0,$$

where α, β, γ are not all zero, is called a (projective) line.

The tangent line *to a projective curve C in $\mathbf{P}_2$ defined by a homogeneous polynomial $P(x, y, z)$ at a nonsingular point $[a, b, c] \in C$ is the line*

$$\frac{\partial P}{\partial x}(a, b, c)x + \frac{\partial P}{\partial y}(a, b, c)y + \frac{\partial P}{\partial z}(a, b, c)z = 0.$$

We give a projective curve C in $\mathbf{P}_2$ the topology it inherits as a subset of $\mathbf{P}_2$ (cf. [Sutherland 75] p.51).

Lemma 2.30 *A projective curve*

$$C = \{[x, y, z] \in \mathbf{P}_2 \ : \ P(x, y, z) = 0\}$$

in $\mathbf{P}_2$ is compact and Hausdorff.

Proof. In order to show that C is compact, by 2.12(iv) and proposition 2.18 it suffices to show that C is a closed subset of $\mathbf{P}_2$. By 2.16(i) this happens if and only if

$$\Pi^{-1}(C) = \{(x, y, z) \in \mathbf{C}^3 - \{0\} \ : \ P(x, y, z) = 0\}$$

is a closed subset of $\mathbf{C}^{n+1} - \{0\}$. This is true since polynomials are continuous.

Any subset of a Hausdorff space is Hausdorff so the result follows from 2.23. $\square$

2.4   Affine and projective curves

Complex algebraic curves

$$C = \{(x, y) \in \mathbf{C}^2 \ : \ Q(x, y) = 0\}$$

in $\mathbf{C}^2$ are often called *affine* curves to distinguish them from *projective* curves

$$\tilde{C} = \{[x, y, z] \in \mathbf{P}_2 \ : \ P(x, y, z) = 0\}.$$

Although different, affine and projective curves are closely related. From an affine curve C one can obtain a projective curve $\tilde{C}$ by adding "points at infinity".

Recall from §2.2 that we can identify $\mathbf{C}^2$ with the open subset

$$U = \{[x, y, z] \in \mathbf{P}_2 \; : \; z \neq 0\}$$

of $\mathbf{P}_2$ via the homeomorphism $\phi : U \to \mathbf{C}^2$ defined by

$$\phi[x, y, z] = \left(\frac{x}{z}, \frac{y}{z}\right)$$

with inverse

$$(x, y) \mapsto [x, y, 1].$$

The complement of U in $\mathbf{P}_2$ is the projective line defined by $z = 0$ which we can identify with $\mathbf{P}_1$ via the map

$$[x, y, 0] \mapsto [x, y].$$

In other words $\mathbf{P}_2$ is the disjoint union of a copy of $\mathbf{C}^2$ and a copy of $\mathbf{P}_1$ which we think of as "at infinity".

Let $P(x, y, z)$ be a nonconstant homogeneous polynomial of degree d. Under the identification of U with $\mathbf{C}^2$ just described, the intersection with U of the projective curve $\tilde{C}$ defined by P is the affine curve in $\mathbf{C}^2$ defined by the inhomogeneous polynomial in two variables

$$P(x, y, 1).$$

This polynomial has degree d provided that z is not a factor of $P(x, y, z)$ (i.e. provided that $\tilde{C}$ does not contain the line $z = 0$).

Conversely if $Q(x, y)$ is an inhomogeneous polynomial of degree d in two variables x and y, say

$$Q(x, y) = \sum_{r+s \leq d} a_{r,s} x^r y^s$$

then the affine curve C defined by $Q(x, y)$ is the intersection of U (identified with $\mathbf{C}^2$) with the projective curve $\tilde{C}$ in $\mathbf{P}_2$ defined by the homogeneous polynomial

$$z^d Q\left(\frac{x}{z}, \frac{y}{z}\right) = \sum_{r+s \leq d} a_{r,s} x^r y^s z^{d-r-s}.$$

The intersection of this projective curve with the line at infinity $z = 0$ is the set of points

$$\{[x, y, 0] \in \mathbf{P}_2 \; : \; \sum_{0 \leq r \leq d} a_{r, d-r} x^r y^{d-r} = 0\}.$$

By lemma 2.8 the homogeneous polynomial

$$\sum_{0 \leq r \leq d} a_{r,d-r} x^r y^{d-r}$$

can be factorised as a product of linear factors

$$\prod_{1 \leq i \leq d} (\alpha_i x + \beta_i y).$$

The lines defined by

$$\alpha_i x + \beta_i y = 0$$

are by definition the *asymptotes* to the curve in $\mathbf{C}^2$ defined by Q. These lines correspond to points $[-\beta_i, \alpha_i]$ in $\mathbf{P}_1$; when $\mathbf{P}_1$ is identified with the line $z = 0$ in $\mathbf{P}_2$ these points are precisely the points of $\tilde{C} - C$.

In this way we get a bijective correspondence between affine curves C in $\mathbf{C}^2$ and projective curves $\tilde{C}$ in $\mathbf{P}_2$ not containing the line at infinity $z = 0$.

If $\tilde{C}$ is nonsingular then so is C but the converse is not necessarily true: $\tilde{C}$ may have singular points at infinity even when C is nonsingular. More precisely we have the following result.

Lemma 2.31 *Let $[a, b, c]$ be a point of the projective curve*

$$\tilde{C} = \{[x, y, z] \in \mathbf{P}_2 \ : \ P(x, y, z) = 0\}.$$

If $c \neq 0$ then $[a, b, c]$ is a nonsingular point of $\tilde{C}$ if and only if $\left(\frac{a}{c}, \frac{b}{c}\right)$ is a nonsingular point of the affine curve

$$C = \{(x, y) \in \mathbf{C}^2 \ : \ P(x, y, 1) = 0\}.$$

Moreover the intersection of $\mathbf{C}^2$ identified with

$$U = \{[x, y, z] \in \mathbf{C}^2 \ : \ z \neq 0\}$$

and the projective tangent line at $[a, b, c]$ to $\tilde{C}$ in $\mathbf{P}_2$ is the tangent line at $\left(\frac{a}{c}, \frac{b}{c}\right)$ to C in $\mathbf{C}^2$.

Proof. For the proof we need

Lemma 2.32 (Euler's Relation) *If $R(x, y, z)$ is a homogeneous polynomial of degree m then*

$$x\frac{\partial R}{\partial x}(x, y, z) + y\frac{\partial R}{\partial y}(x, y, z) + z\frac{\partial R}{\partial z}(x, y, z) = mR(x, y, z).$$

Given this lemma we can complete the proof of lemma 2.31. The point $\left(\frac{a}{c}, \frac{b}{c}\right)$ is a singular point of C if and only if

$$P\left(\frac{a}{c}, \frac{b}{c}, 1\right) = 0 = \frac{\partial P}{\partial x}\left(\frac{a}{c}, \frac{b}{c}, 1\right) = \frac{\partial P}{\partial y}\left(\frac{a}{c}, \frac{b}{c}, 1\right).$$

Since $P(x, y, z)$ and its partial derivatives are homogeneous and $c \neq 0$ this happens if and only if

$$P(a, b, c) = 0 = \frac{\partial P}{\partial x}(a, b, c) = \frac{\partial P}{\partial y}(a, b, c),$$

and Euler's relation tells us that this happens if and only if

$$P(a, b, c) = 0 = \frac{\partial P}{\partial x}(a, b, c) = \frac{\partial P}{\partial y}(a, b, c) = \frac{\partial P}{\partial z}(a, b, c),$$

i.e. if and only if $[a, b, c]$ is a singular point of $\tilde{C}$.

The intersection of $\mathbf{C}^2$ identified with U and the projective tangent line

$$x\frac{\partial P}{\partial x}(a, b, c) + y\frac{\partial P}{\partial y}(a, b, c) + z\frac{\partial P}{\partial z}(a, b, c) = 0$$

is the line in $\mathbf{C}^2$ defined by

$$x\frac{\partial P}{\partial x}(a, b, c) + y\frac{\partial P}{\partial y}(a, b, c) + \frac{\partial P}{\partial z}(a, b, c) = 0.$$

By the homogeneity of the partial derivatives and Euler's relation again this is precisely the tangent line

$$(x - \frac{a}{c})\frac{\partial P}{\partial x}\left(\frac{a}{c}, \frac{b}{c}, 1\right) + (y - \frac{b}{c})\frac{\partial P}{\partial y}\left(\frac{a}{c}, \frac{b}{c}, 1\right) = 0$$

to C in $\mathbf{C}^2$ at $\left(\frac{a}{c}, \frac{b}{c}\right)$.

It remains to prove Euler's relation.

Proof of lemma 2.32. Euler's relation is obtained by differentiating the identity

$$R(\lambda x, \lambda y, \lambda z) = \lambda^m R(x, y, z)$$

with respect to λ and then setting $\lambda = 1$. $\square$

Remark 2.33 In a similar way definitions such as those given in §2.1 of the multiplicity and tangent lines of singular points on affine curves can be modified to apply to projective curves so that the definitions respect the identification of projective curves with affine curves together with extra "points at infinity" (cf. exercise 2.8).

2.5   Exercises

2.1. Show that the subset of $\mathbf{C}^2$ consisting of points of the form

$$(t^2, t^3 + 1), \quad t \in \mathbf{C}$$

is a complex algebraic curve.

2.2. Find the singular points, and the tangent lines at the singular points, of each of the following curves in $\mathbf{C}^2$:

(a). $y^3 - y^2 + x^3 - x^2 + 3y^2 x + 3x^2 y + 2xy = 0$.
(b). $x^4 + y^4 - x^2 y^2 = 0$.
(c). $y^2 = x^3 - x$.

2.3. If $P(x, y)$ is a polynomial of degree d and a and b are complex numbers, show that $P(x, y)$ is given by

$$\sum_{0 \leq i+j \leq d} \frac{1}{i! j!} (x - a)^i (y - b)^j \frac{\partial^{i+j} P}{\partial x^i \partial y^j}(a, b).$$

2.4. Let (a, b) be a singular point of an affine curve C in $\mathbf{C}^2$ defined by a polynomial $P(x, y)$. Show that (a, b) is an ordinary double point if and only if

$$\left(\frac{\partial^2 P}{\partial x \partial y} \right)^2 \neq \left(\frac{\partial^2 P}{\partial x^2} \right) \left(\frac{\partial^2 P}{\partial y^2} \right)$$

at the point (a, b).

2.5. Let C be an affine curve defined by a polynomial $P(x, y)$ of degree d. Show that if (a, b) is a point of multiplicity d in C then $P(x, y)$ is a product of d linear factors, so C is the union of d lines through (a, b).

2.6. (a). Show that the union of finitely many affine curves in $\mathbf{C}^2$ is an affine curve.

(b). Show that the union of finitely many projective curves in the projective plane is a projective curve.

2.7. Show that a complex algebraic curve in $\mathbf{C}^2$ is never compact. [Hint: recall that a subset of $\mathbf{C}^2$ is compact if and only if it is closed and bounded. Show that if $P(x, y)$ is any nonconstant polynomial with complex coefficients then for all but at most finitely many values of $a \in \mathbf{C}$ there exists $b \in \mathbf{C}$ such that $P(a, b) = 0$].

2.8. The multiplicity at a point $[a, b, c]$ of a projective curve C defined by $P(x, y, z) = 0$ is the smallest integer m such that

$$\frac{\partial^m P}{\partial x^i y^j z^k}(a, b, c) \neq 0$$

for some i, j, k such that $i + j + k = m$. Find the singular points and the multiplicities of the singular points of the following projective curves.

(a). $xy^4 + yz^4 + xz^4 = 0$.
(b). $x^2y^3 + x^2z^3 + y^2z^3 = 0$.
(c). $y^2z = x(x - z)(x - \lambda z)$, $\lambda \in \mathbf{C}$.
(d). $x^n + y^n + z^n = 0$, $n > 0$.

2.9. For which values of $\lambda \in \mathbf{C}$ are the following projective curves in $\mathbf{P}_2$ nonsingular? Describe the singularities when they exist.

(a). $x^3 + y^3 + z^3 + 3\lambda xyz = 0$.
(b). $x^3 + y^3 + z^3 + \lambda(x + y + z)^3 = 0$.

2.10. Suppose nine distinct points in $\mathbf{P}_2$ are given, and that they do not all lie on any line in $\mathbf{P}_2$. Suppose that any straight line which passes through two of these points also passes through a third. Show that there is a projective transformation taking these points to the points

$$
\begin{array}{ccc}
[0,1,-1] & [-1,0,1] & [1,-1,0] \\
[0,1,\alpha] & [\alpha,0,1] & [1,\alpha,0] \\
[0,\alpha,1] & [1,0,\alpha] & [\alpha,1,0]
\end{array}
$$

for some $\alpha \in \mathbf{C}$. Show that $\alpha^2 - \alpha + 1 = 0$. Show also that a projective curve of degree 3 passes through these nine points if and only if it is defined by a polynomial of the form

$$x^3 + y^3 + z^3 + 3\lambda xyz$$

for some $\lambda \in \mathbf{C} \cup \{\infty\}$, and it is singular precisely when

$$\lambda \in \{\infty, -1, \alpha, \overline{\alpha}\},$$

in which case it is the union of three lines in $\mathbf{P}_2$.

2.11. Show that the complex line in $\mathbf{P}_2$ through the points $[0,1,1]$ and $[t,0,1]$ meets the projective curve C defined by

$$x^2 + y^2 = z^2$$

in the two points $[0,1,1]$ and $[2t, t^2 - 1, t^2 + 1]$. Show that there is a bijection from the complex line defined by $y = 0$ to C given by

$$[t,0,1] \mapsto [2t, t^2 - 1, t^2 + 1]$$

and

$$[1,0,0] \mapsto [0,1,1].$$

Deduce that the complex solutions to Pythagoras' equation

$$x^2 + y^2 = z^2$$

are

$$x = 2\lambda\mu, \quad y = \lambda^2 - \mu^2, \quad z = \lambda^2 + \mu^2$$

with $\lambda, \mu \in \mathbf{C}$. Show that the real solutions are

$$x = 2\lambda\mu, \quad y = \lambda^2 - \mu^2, \quad z = \pm(\lambda^2 + \mu^2)$$

with $\lambda, \mu \in \mathbf{R}$, and the integer solutions are

$$x = 2\lambda\mu\nu, \quad y = (\lambda^2 - \mu^2)\nu, \quad z = (\lambda^2 + \mu^2)\nu$$

where λ and μ are coprime integers, not both odd, and $\nu \in \mathbf{Z}$, or

$$x = \lambda\mu\nu, \quad y = \frac{1}{2}(\lambda^2 - \mu^2)\nu, \quad z = \frac{1}{2}(\lambda^2 + \mu^2)\nu$$

where λ and μ are coprime odd integers and $\nu \in \mathbf{Z}$.

2.12. Let C be a nonsingular projective curve of degree two in $\mathbf{P}_2$ defined by a polynomial with *rational* coefficients. Use the following steps to obtain an algorithm which decides whether or not C has any *rational points*, i.e. whether there is a point of C which can be represented by rational homogeneous coordinates (or equivalently by integral homogeneous coordinates).

(a). Use the theory of the diagonalisation of quadratic forms to show that there is a projective transformation defined by a matrix with rational coefficients taking C to the curve defined by

$$ax^2 + by^2 = z^2$$

for some $a, b \in \mathbf{Q} - \{0\}$.

(b). Show that by making an additional diagonal transformation we can assume that a and b are integers with no square factors, i.e. each is a product of distinct primes. Show that we may also assume that $|a| \geq |b|$.

(c). Show that if C has a rational point then b is a square modulo p for every prime p dividing a. Deduce from the Chinese remainder theorem that b is a square modulo a, so there are integers m, a_1 such that $|m| \leq |a|/2$ and

$$m^2 = b + aa_1.$$

(d). Show that if $m^2 = b + aa_1$ and

$$ax^2 + by^2 = z^2$$

then

$$a_1(z^2 - by^2)^2 + b(my - z)^2 x^2 = (mz - by)^2 x^2.$$

Deduce that C has a rational point if and only if the same is true of the curve defined by

$$a_1 x^2 + b y^2 = z^2.$$

(e). Show that if $|\,a\,| > 1$ then $|\,a_1\,| < |\,a\,|$, and thus the problem is reduced to one of the same form in which $|\,a\,| + |\,b\,|$ is smaller. Deduce that the argument can be repeated until either b fails to be a square modulo a or we reach the situation

$$|\,a\,| = |\,b\,| = 1,$$

in which case the curve has a rational point if and only if at least one of a and b is positive.

2.13. Suppose that C is a nonsingular projective curve of degree two in $\mathbf{P}_2$ defined by a polynomial with rational coefficients and that C has a rational point p (cf. exercise 2.12). Show that there is a bijection from the set of rational points of C to the set of rational points of a projective line given by stereographic projection (cf. exercise 2.11). [If L is any line in $\mathbf{P}_2$ defined by a linear polynomial with rational coefficients and which does not contain p, then stereographic projection from p onto L associates to any point $q \in C$ the unique point of intersection of L with the line through p and q if $p \neq q$, or the tangent line to C at p if $p = q$].

2.14. Let p be any odd prime. If $\lambda \in \mathbf{Z}$ we can reduce λ modulo p and define the reduction modulo p of the projective curve C of degree three defined by

$$y^2 z = x(x - z)(x - \lambda z)$$

to be the subset of the projective plane $\mathbf{P}((\mathbf{F}_p)^3)$ over the finite field $\mathbf{F}_p = \mathbf{Z}/p\mathbf{Z}$ defined by the corresponding equation. Show that for given $x \in \mathbf{F}_p$ the equation

$$y^2 = x(x - 1)(x - \lambda)$$

has one solution if $x = 0, 1$ or λ, two solutions if

$$(x(x - 1)(x - \lambda))^{(p-1)/2} \equiv 1 \ (mod \ p)$$

and no solutions otherwise, in which case

$$(x(x - 1)(x - \lambda))^{(p-1)/2} \equiv -1 \ (mod \ p).$$

[Hint: recall that the multiplicative group of any finite field is cyclic]. Deduce that the number of points in the reduction of C modulo p is

$$1 + \sum_{x \in \mathbf{F}_p} (1 + (x(x - 1)(x - \lambda))^{(p-1)/2}.$$

By expanding $(x(x-1)(x-\lambda))^{(p-1)/2}$ as a polynomial in x and using the formulas

$$\sum_{x \in \mathbf{F}_p} x^l \equiv \begin{cases} 0 & (mod \ p) \quad \text{if} \ \ p-1 \nmid l \\ -1 & (mod \ p) \quad \text{if} \ \ p-1 \mid l \end{cases}$$

show that the number of points in the reduction of C modulo p is

$$1 + p + (-1)^{(p-1)/2} \sum_{r=0}^{(p-1)/2} \binom{\frac{p-1}{2}}{r} \lambda^r.$$

Chapter 3

Algebraic properties

In this chapter we shall study some of the algebraic properties of complex algebraic curves. In §3.1 we shall investigate the way in which two complex projective curves in $\mathbf{P}_2$ can meet each other. In §3.2 we shall study the "inflection points" of a complex projective curve and show that every nonsingular projective curve of degree three is equivalent under a projective transformation to one defined by

$$y^2 z = x(x - z)(x - \lambda z)$$

for some $\lambda \in \mathbf{C} - \{0, 1\}$.

3.1 Bézout's theorem

In this section we shall study the way in which two projective curves C and D in $\mathbf{P}_2$ can intersect. We shall see that C and D always intersect in at least one point, and that if C and D have no common component then they intersect in at most nm points where n is the degree of C and m is the degree of D. We shall also see that C and D meet in *exactly* nm points if every point of $C \cap D$ is a nonsingular point of both C and D and the tangent lines to C and D at such points are distinct. These results are different cases of the theorem usually known by the name of an eighteenth century French mathematician Bézout[1].

We shall be able to prove a more general result about the number of points of intersection of C and D once we have defined the intersection multiplicity $I_p(C, D)$ of C and D at a point p. This is defined to be infinity if p lies on a common component of C and D, and otherwise it is a nonnegative integer

---

[1]The reasons for giving Bézout's name to this theorem are not entirely clear, since although Bézout gave a proof of the theorem it was neither correct nor the first proof to be given.

which is zero precisely when p does not belong to $C \cap D$. We shall show that $I_p(C, D)$ is one if and only if p is a nonsingular point of both C and D and the tangent lines to C and D at p are distinct. The strongest form of Bézout's theorem can then be expressed as follows.

Theorem 3.1 (Bézout's theorem) *If C and D are two projective curves of degrees n and m in $\mathbf{P}_2$ which have no common component then they have precisely nm points of intersection counting multiplicities; i.e.*

$$\sum_{p \in C \cap D} I_p(C, D) = nm.$$

The crucial ingredient in our proof of Bézout's theorem is the concept of a *resultant* (sometimes called an *eliminant*).

Definition 3.2 *Let*

$$P(x) = a_0 + a_1 x + \ldots + a_n x^n,$$

where $a_0, \ldots, a_n \in \mathbf{C}, a_n \neq 0$, and

$$Q(x) = b_0 + b_1 x + \ldots + b_m x^m,$$

where $b_0, \ldots, b_m \in \mathbf{C}, b_m \neq 0$, be polynomials of degrees n and m in x. The resultant $\mathcal{R}_{P,Q}$ of P and Q is the determinant of the $m + n$ by $m + n$ matrix

$$\begin{pmatrix} a_0 & a_1 & \ldots & a_n & 0 & 0 & & \ldots & 0 \\ 0 & a_0 & a_1 & \ldots & a_n & 0 & & \ldots & 0 \\ \cdot & & & \cdots & & & & & \cdot \\ & & & \cdots & & & & & \\ \cdot & & & \cdots & & & & & \cdot \\ 0 & 0 & \ldots & 0 & a_0 & a_1 & & \ldots & a_n \\ b_0 & b_1 & & \ldots & & b_m & 0 & \ldots & 0 \\ 0 & b_0 & b_1 & & \ldots & & b_m & 0 & \ldots & 0 \\ \cdot & & & \cdots & & & & & \cdot \\ 0 & \ldots & 0 & b_0 & b_1 & & & \ldots & b_m \end{pmatrix}.$$

If

$$P(x, y, z) = a_0(y, z) + a_1(y, z)x + \ldots + a_n(y, z)x^n$$

and

$$Q(x, y, z) = b_0(y, z) + b_1(y, z)x + \ldots + b_m(y, z)x^m$$

are polynomials in three variables x, y, z then the resultant

$$\mathcal{R}_{P,Q}(y, z)$$

of P and Q with respect to x is defined as a determinant in exactly the same way as $\mathcal{R}_{P,Q}$ was defined above but with $a_i(y,z)$ and $b_j(y,z)$ replacing a_i and b_j for $0 \leq i \leq n$ and $0 \leq j \leq m$. Note that $\mathcal{R}_{P,Q}(y,z)$ is a polynomial in y and z whose value when $y = b$ and $z = c$ is the resultant of the polynomials $P(x,b,c)$ and $Q(x,b,c)$ in x, provided that $a_n(b,c)$ and $b_m(b,c)$ are nonzero[†].

The following lemmas, whose proofs will be postponed to the end of the section, indicate why resultants are useful in the study of the intersection of projective curves in $\mathbf{P}_2$.

Lemma 3.3 *Let $P(x)$ and $Q(x)$ be polynomials in x. Then $P(x)$ and $Q(x)$ have a nonconstant common factor if and only if*

$$\mathcal{R}_{P,Q} = 0.$$

Lemma 3.4 *Let $P(x,y,z)$ and $Q(x,y,z)$ be nonconstant homogeneous polynomials in x,y,z such that*

$$P(1,0,0) \neq 0 \neq Q(1,0,0).$$

Then $P(x,y,z)$ and $Q(x,y,z)$ have a nonconstant homogeneous common factor if and only if the polynomial $\mathcal{R}_{P,Q}(y,z)$ in y and z is identically zero.

Remark 3.5 The reason for the requirement that

$$P(1,0,0) \neq 0 \neq Q(1,0,0)$$

in this lemma is to ensure that the polynomials $P(x,y,z)$ and $Q(x,y,z)$ have the same degree when regarded as polynomials in x with coefficients in the ring $\mathbf{C}[y,z]$ of polynomials in y and z as they do when regarded as polynomials in x,y and z together.

Lemma 3.6 *If*
$$P(x) = (x - \lambda_1)...(x - \lambda_n)$$

and

$$Q(x) = (x - \mu_1)...(x - \mu_m)$$

where $\lambda_1,...\lambda_n, \mu_1,...,\mu_m$ are complex numbers, then

$$\mathcal{R}_{P,Q} = \prod_{1 \leq i \leq n, 1 \leq j \leq m} (\mu_j - \lambda_i).$$

In particular

$$R_{P,QR} = \mathcal{R}_{P,Q} R_{P,R}$$

when P, Q and R are polynomials in x. The corresponding result is also true when P, Q and R are polynomials in x, y and z.

Lemma 3.7 *Let $P(x, y, z)$ and $Q(x, y, z)$ be homogeneous polynomials of degrees n and m in x, y, z. Then the resultant $\mathcal{R}_{P,Q}(y, z)$ is a homogeneous polynomial of degree nm in y and z.*

Assuming these lemmas, we can prove

Theorem 3.8 *Any two projective curves C and D in $\mathbf{P}_2$ intersect in at least one point.*

Proof. Let C and D be defined by homogeneous polynomials $P(x, y, z)$ and $Q(x, y, z)$ of degrees n and m. By lemma 3.7 the resultant $\mathcal{R}_{P,Q}(y, z)$ is a homogeneous polynomial of degree nm in y and z. Therefore by lemma 2.8 $\mathcal{R}_{P,Q}(y, z)$ is either identically zero or it is the product of nm linear factors $bz - cy$ with b, c complex numbers, not both zero. In either case there exists $(b, c) \in \mathbf{C}^2 - \{0\}$ such that $\mathcal{R}_{P,Q}(y, z)$ vanishes when $y = b$ and $z = c$. This means that the resultant of the polynomials $P(x, b, c)$ and $Q(x, b, c)$ in x is zero, so by lemma 3.3 these polynomials have a common root $a \in \mathbf{C}$. Then

$$P(a, b, c) = 0 = Q(a, b, c)$$

so $[a, b, c] \in C \cap D$. □

We can also prove

Theorem 3.9 (Weak form of Bézout's theorem). *If two projective curves C and D in $\mathbf{P}_2$ of degrees n and m have no common component then they intersect in at most nm points.*

Proof. Suppose that C and D have at least $nm + 1$ points of intersection. We shall show that C and D have a common component. Choose any set S of $nm + 1$ distinct points in $C \cap D$. Then we can choose a point in $\mathbf{P}_2$ which does not lie on C or on D or on any of the finitely many lines in $\mathbf{P}_2$ passing through two distinct points of S. By applying a projective transformation we may assume that this point is $[1, 0, 0]$. Then the curves C and D are defined by homogeneous polynomials $P(x, y, z)$ and $Q(x, y, z)$ of degrees n and m such that

$$P(1, 0, 0) \neq 0 \neq Q(1, 0, 0)$$

because $[1, 0, 0]$ does not belong to $C \cup D$. By lemma 3.7 the resultant $\mathcal{R}_{P,Q}(y, z)$ of P and Q with respect to x is a homogeneous polynomial of degree nm in y and z, so if $\mathcal{R}_{P,Q}(y, z)$ is not identically zero then it is the product of nm linear factors of the form $bz - cy$ where $(b, c) \in \mathbf{C}^2 - \{0\}$. Moreover if $(b, c) \in \mathbf{C}^2 - \{0\}$ then $bz - cy$ is a factor of $\mathcal{R}_{P,Q}(y, z)$ if and only if the resultant of the polynomials $P(x, b, c)$ and $Q(x, b, c)$ in x vanishes, and hence by lemma 3.3 if and only if there is some $a \in \mathbf{C}$ such that

$$P(a, b, c) = 0 = Q(a, b, c).$$

But if $[a, b, c] \in S$ then $P(a, b, c) = 0 = Q(a, b, c)$ and b and c are not both zero (because $[1, 0, 0]$ does not belong to S) so $bz - cy$ is a factor of $\mathcal{R}_{P,Q}(y, z)$. Moreover if $[\alpha, \beta, \gamma] \in S$ is distinct from $[a, b, c] \in S$ then $\beta z - \gamma y$ is not a scalar multiple of $bz - cy$, because otherwise $[a, b, c], [\alpha, \beta, \gamma]$ and $[1, 0, 0]$ would all lie on the line in $\mathbf{P}_2$ defined by

$$bz = cy$$

which contradicts the assumption on $[1, 0, 0]$. This shows that $\mathcal{R}_{P,Q}(y, z)$ has at least $nm + 1$ distinct linear factors, so it must be identically zero. By lemma 3.4 this implies that C and D have a common component. $\square$

The next results are all applications of these two theorems.

Corollary 3.10 *(i) A nonsingular projective curve C in $\mathbf{P}_2$ is irreducible.*
(ii) An irreducible projective curve C in $\mathbf{P}_2$ has at most finitely many singular points.

Proof. (i) Let

$$C = \{[x, y, z] \in \mathbf{P}_2 \ : \ P(x, y, z)Q(x, y, z) = 0\}$$

be a reducible projective curve in $\mathbf{P}_2$. Then by theorem 3.8 there is at least one point $[a, b, c] \in \mathbf{P}_2$ such that

$$P(a, b, c) = 0 = Q(a, b, c).$$

It is easy to check that $[a, b, c]$ is a singular point of C.

(ii) Let C be defined by a homogeneous polynomial $P(x, y, z)$ of degree n. Without loss of generality we may assume that $[1, 0, 0]$ does not belong to C so the coefficient $P(1, 0, 0)$ of x^n in $P(x, y, z)$ is not equal to zero. This ensures that

$$Q(x, y, z) = (\partial P/\partial x)(x, y, z)$$

is a homogeneous polynomial of degree $n - 1$ which is not identically zero and hence defines a curve D in $\mathbf{P}_2$. Since C is irreducible and the degree of D is strictly less than the degree of C the curves C and D can have no common component. Hence by Bézout's theorem C and D intersect in at most $n(n - 1)$ points. As every singular point of C lies in $C \cap D$ the result follows. $\square$

Definition 3.11 *A conic is a curve of degree two in $\mathbf{C}^2$ or $\mathbf{P}_2$.*

Corollary 3.12 *Any irreducible projective conic C in $\mathbf{P}_2$ is equivalent under a projective transformation to the conic*

$$x^2 = yz.$$

and in particular is nonsingular.

Proof. By corollary 3.10, C has at most finitely many singular points. Thus by applying a suitable projective transformation we may assume that $[0,1,0]$ is a nonsingular point of C and that the tangent line to C at $[0,1,0]$ is the line $z = 0$. Then C must be defined by a polynomial of the form

$$ayz + bx^2 + cxz + dz^2$$

for some complex numbers a, b, c, d. Since C is irreducible, a and b are both nonzero. The projective transformation

$$[x, y, z] \mapsto [\sqrt{b}x, ay + cx + dz, -z]$$

takes C to the conic $x^2 = yz$. Since this conic is nonsingular it follows that C is nonsingular as well. $\square$

Remark 3.13 Let C be the nonsingular conic defined by

$$x^2 = yz$$

in $\mathbf{P}_2$. There is a homeomorphism $f : \mathbf{P}_1 \to C$ given by

$$f[x, y] = [xy, y^2, x^2]$$

with inverse $g : C \to \mathbf{P}_1$ given by

$$g[x, y, z] = \begin{cases} [x, y] & \text{if } y \neq 0 \\ [z, x] & \text{if } z \neq 0. \end{cases}$$

(Note that if $[x, y, z] \in C$ then $x^2 = yz$ so if $y \neq 0 \neq z$ then $x \neq 0$ and

$$[x, y] = [x^2, xy] = [yz, xy] = [z, x]).$$

Thus by corollary 3.12 any irreducible projective conic in $\mathbf{P}_2$ is homeomorphic to $\mathbf{P}_1$.

The next result is another application of Bézout's theorem.

Proposition 3.14 *If two projective curves C and D of degrees n in $\mathbf{P}_2$ intersect in exactly n^2 points and if exactly nm of these points lie on an irreducible curve E of degree $m < n$ then the remaining $n(n-m)$ points lie on a curve of degree at most $n - m$.*

Proof. Let C, D and E be defined by the homogeneous polynomials

$$P(x, y, z), Q(x, y, z), R(x, y, z)$$

respectively. Choose a point $[a, b, c]$ on E which does not lie on $C \cap D$. Then the curve of degree n defined by

$$\lambda P(x, y, z) + \mu Q(x, y, z)$$

where

$$\lambda = Q(a, b, c), \mu = -P(a, b, c),$$

meets E in at least $nm + 1$ points, namely $[a, b, c]$ and the nm points of $C \cap D$ which lie on E by hypothesis. Then by theorem 3.9 (the weak form of Bézout's theorem) this curve and E must have a common component, which must be E itself because E is irreducible. Thus

$$\lambda P(x, y, z) + \mu Q(x, y, z) = R(x, y, z)S(x, y, z)$$

for some nonconstant homogeneous polynomial $S(x, y, z)$ of degree $n - m$. Hence if $[u, v, w] \in C \cap D$ then either $R(u, v, w) = 0$ or $S(u, v, w) = 0$. Therefore the $n(n - m)$ points of $C \cap D$ which do not lie on E must all lie on the curve defined by $S(x, y, z)$. □

Corollary 3.15 (Pascal's mystic hexagon). *The pairs of opposite sides of a hexagon inscribed in an irreducible conic in* $\mathbf{P}_2$ *meet in three collinear points.*

Remark 3.16 This result is to be interpreted as follows. A hexagon in $\mathbf{P}_2$ is simply determined by six distinct points $p_1, \ldots, p_6$ in $\mathbf{P}_2$ (its vertices) and its sides are the lines in $\mathbf{P}_2$ joining p_1 to p_2, p_2 to p_3, p_3 to p_4, p_4 to p_5, p_5 to p_6 and p_6 and p_1. The side opposite the line joining p_1 to p_2 is the line joining p_4 to p_5, and so on. The hexagon is said to be inscribed in a conic if its vertices lie on the conic. Three points of $\mathbf{P}_2$ are said to be collinear if they all lie on some line in $\mathbf{P}_2$.

Proof of corollary 3.15. Let the successive sides of the hexagon be the lines defined by linear polynomials $L_1, \ldots, L_6$ in x, y, z. The two projective curves of degree three defined by $L_1 L_3 L_5$ and $L_2 L_4 L_6$ intersect in the six vertices of the hexagon and the three points of intersection of the opposite sides of the hexagon. The result now follows immediately from proposition 3.14. □

Remark 3.17 If we assume that the conic is defined by a polynomial with real coefficients and the vertices of the hexagon lie in the subset $\mathbf{R}^2$ of $\mathbf{C}^2 \subseteq \mathbf{P}_2$ then we obtain a theorem about real Euclidean geometry (figure 3.1).

In order to prove the stronger form 3.1 of Bézout's theorem, we must first define the intersection multiplicity $I_p(C, D)$ at a point $p = [a, b, c]$ of two

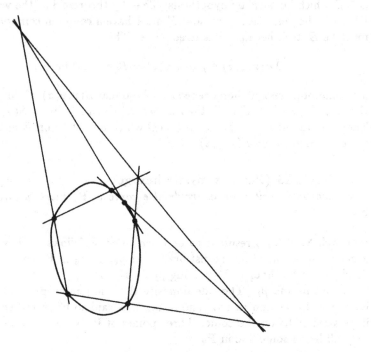

Figure 3.1: Pascal's mystic hexagon

curves C and D in $\mathbf{P}_2$. We shall define the intersection multiplicity by using the resultant of the polynomials $f(x,y,z)$ and $g(x,y,z)$ defining C and D in a suitable coordinate system. In order to show that the definition is independent of the choice of coordinates we show that it is uniquely determined by the properties listed in the following theorem.

Theorem 3.18 *There is a unique intersection multiplicity $I_p(C,D)$ defined for all projective curves C and D in $\mathbf{P}_2$ satisfying the following properties (i)–(vi).*

(i). $I_p(C,D) = I_p(D,C)$.

(ii). $I_p(C,D) = \infty$ if p lies on a common component of C and D, and otherwise $I_p(C,D)$ is a nonnegative integer.

(iii). $I_p(C,D) = 0$ if and only if $p \notin C \cap D$.

(iv). Two distinct lines meet with intersection multiplicity one at their unique point of intersection.

(v). If C_1 and C_2 are defined by homogeneous polynomials $P_1(x,y,z)$ and $P_2(x,y,z)$ and C is defined by

$$P(x,y,z) = P_1(x,y,z)P_2(x,y,z)$$

then

$$I_p(C,D) = I_p(C_1,D) + I_p(C_2,D).$$

(vi). If C and D are defined by homogeneous polynomials $P(x,y,z)$ and $Q(x,y,z)$ of degrees n and m and E is defined by $PR+Q$ where $R(x,y,z)$ is homogeneous of degree $m-n$ then

$$I_p(C,D) = I_p(C,E).$$

Moreover if C and D have no common component and we choose projective coordinates so that the conditions

(a). $[1,0,0]$ does not belong to $C \cup D$;

(b). $[1,0,0]$ does not lie on any line containing two distinct points of $C \cap D$;

(c). $[1,0,0]$ does not lie on the tangent line to C or D at any point of $C \cap D$;

are satisfied then the intersection multiplicity $I_p(C,D)$ of C and D at any

$$p = [a,b,c] \in C \cap D$$

is the largest integer k such that $(bz-cy)^k$ divides the resultant $\mathcal{R}_{P,Q}(y,z)$.

Remark 3.19 A careful study of the uniqueness proof in this theorem will reveal that for the proof to be valid we need to allow curves to have multiple components; i.e. the polynomials defining them may have repeated factors (see remarks 2.3 and 2.25). In fact all the arguments of this chapter apply without alteration to curves with multiple components.

Proof of theorem 3.18. To simplify notation in this proof we shall write $I_p(P, Q)$ instead of $I_p(C, D)$ when $P(x, y, z)$ and $Q(x, y, z)$ are homogeneous polynomials defining the curves C and D.

First we shall show that the intersection multiplicity $I_p(P, Q)$ can be calculated using only the conditions (i)–(vi), so that these conditions determine $I_p(P, Q)$ completely. Since the conditions are independent of the choice of coordinates we may assume that $p = [0, 0, 1]$. Moreover we may assume that P and Q are irreducible by (i) and (v), that $I_p(P, Q)$ is finite by (ii), and that $I_p(P, Q) = k > 0$ by (iii). Finally by induction on k we may assume that any intersection multiplicity strictly less than k can be calculated using only the conditions (i)–(vi).

Consider the polynomials $P(x, 0, 1)$ and $Q(x, 0, 1)$ in x; let them have degrees r and s respectively. By (i) we may assume that $r \leq s$. There are two cases to consider.

Case 1: $r = 0$. In this case $P(x, 0, 1)$ is constant and hence zero because $P(0, 0, 1) = 0$. Since $P(x, y, z)$ is a homogeneous polynomial it follows that $P(x, 0, z)$ is identically zero, and hence that

$$P(x, y, z) = y R(x, y, z)$$

for some homogeneous polynomial $R(x, y, z)$. Moreover we can write

$$Q(x, y, z) = Q(x, 0, z) + y S(x, y, z) = x^q T(x, z) + y S(x, y, z)$$

for some homogeneous polynomials $T(x, z)$ and $S(x, y, z)$ such that $T(0, 1)$ is not equal to zero, and some integer q which is *positive* since $Q(0, 0, 1) = 0$. Note that the condition $T(0, 1) \neq 0$ means that the point $p = [0, 0, 1]$ does not lie on the curve defined by $T(x, z) = 0$, and hence by (iii) that

$$I_p(y, T(x, z)) = 0,$$

whereas from (iv) we have

$$I_p(y, x) = 1.$$

Putting this information together, from (v) we obtain

$$I_p(P, Q) = I_p(R, Q) + I_p(y, Q),$$

from (vi) we get

$$I_p(y, Q) = I_p(y, x^q T(x, z))$$

and by repeated use of (v) and (ii)

$$I_p(y, x^q T(x, z)) = q I_p(y, x) + I_p(y, T(x, z)) = q.$$

Hence

$$I_p(P, Q) = I_p(R, Q) + q,$$

and since $q > 0$ our inductive hypothesis tells us that $I_p(R, Q)$ can be calculated using only the conditions (i)–(vi).

Case 2: $r > 0$. In this case we can multiply $P(x, y, z)$ and $Q(x, y, z)$ by constants to make the polynomials $P(x, 0, 1)$ and $Q(x, 0, 1)$ in x monic. If n and m are the degrees of $P(x, y, z)$ and $Q(x, y, z)$ consider the polynomial

$$S(x, y, z) = z^{n+s-r} Q(x, y, z) - x^{s-r} z^m P(x, y, z).$$

This is constructed to be a homogeneous polynomial in x, y, z such that the polynomial

$$S(x, 0, 1) = Q(x, 0, 1) - x^{s-r} P(x, 0, 1)$$

in x has degree t strictly less than s. Note that $S(x, y, z)$ is not identically zero since by assumption $P(x, y, z)$ and $Q(x, y, z)$ are irreducible and distinct. Moreover by (i), (v) and (vi)

$$I_p(P, S) = I_p(P, z^{n+s-r} Q) = I_p(P, Q).$$

Now replace P and Q by P and S (or by S and P if $t < r$). After repeating this process a finite number of times we reach the situation of Case 1.

This completes the uniqueness part of the proof. To prove existence, let us define the intersection multiplicity $I_p(C, D)$ as follows.

- If p lies on a common component of C and D then $I_p(C, D) = \infty$.

- If p does not belong to $C \cap D$ then $I_p(C, D) = 0$.

- If p belongs to $C \cap D$ but does not lie on a common component of C and D, first remove any common components from C and D and then choose coordinates such that the conditions (a)–(c) are satisfied. If $p = [a, b, c]$ in these coordinates then $I_p(C, D)$ is the largest integer k such that $(bz - cy)^k$ divides the resultant $\mathcal{R}_{P,Q}(y, z)$ of P and Q with respect to x.

It remains to show that the conditions (i)–(vi) are now satisfied.

(i) is a direct consequence of the fact that interchanging two rows of a determinant changes its sign and hence

$$\mathcal{R}_{P,Q}(y, z) = \pm \mathcal{R}_{Q,P}(y, z).$$

(ii) follows from the definition and lemma 3.4.

(iii) is easy. If $p = [a, b, c] \in C \cap D$ then the polynomials $P(x, b, c)$ and $Q(x, b, c)$ have a common root a, so by lemma 3.4 the homogeneous polynomial $\mathcal{R}_{P,Q}(y, z)$ vanishes when $y = b$ and $z = c$. Hence it is divisible by $bz - cy$, so $I_p(C, D) > 0$.

(iv) is a straightforward calculation with two-by-two determinants.

(v) follows immediately from lemma 3.6.

(vi) is true because a determinant is unchanged by the addition of a scalar multiple of one row to another. The resultant of P and $PR + Q$ is the determinant of a matrix (s_{ij}) obtained from the matrix (r_{ij}) defining $\mathcal{R}_{P,Q}(y, z)$ by the addition of suitable scalar multiples of the first n rows to the last m rows. More precisely, if

$$R(x, y, z) = \rho_0(y, z) + \rho_1(y, z)x + \dots + \rho_{n-m}(y, z)x^{n-m}$$

then

$$s_{ij} = \begin{cases} r_{ij} & \text{if } i \leq m \\ r_{ij} + \sum_{k=i-m}^{i-n} \rho_{i-n-k} r_{kj} & \text{if } i > m \end{cases}$$

so that

$$\mathcal{R}_{P,PR+Q}(y, z) = \det(s_{ij}) = \det(r_{ij}) = \mathcal{R}_{P,Q}(y, z).$$

This completes the proof of the theorem and the definition of the intersection multiplicity $I_p(C, D)$. $\square$

Remark 3.20 The uniqueness part of this proof shows that the calculation of intersection multiplicities is not difficult. It only involves some simple arithmetic manipulation of polynomials.

Remark 3.21 It is a consequence of (iii) and (v) that the intersection multiplicity $I_p(C, D)$ depends only on those components of C and D which contain p.

We can now prove theorem 3.1, the strong form of Bézout's theorem.

Proof of theorem 3.1. Let C and D be projective curves of degrees n and m in $\mathbf{P}_2$ with no common component. We must prove that the number of points of intersection counting multiplicities is nm, i.e. that

$$\sum_{p \in C \cap D} I_p(C, D) = nm.$$

We may choose coordinates so that the conditions (a)–(c) of theorem 3.18 are satisfied. Let C and D be defined by the homogeneous polynomials $P(x, y, z)$ and $Q(x, y, z)$ in this coordinate system. By lemmas 3.4 and 3.7 the resultant $\mathcal{R}_{P,Q}(y, z)$ is a homogeneous polynomial of degree nm in the two variables y and z, not identically zero, so by lemma 2.8 it can be expressed as a product of nm linear factors,

$$\mathcal{R}_{P,Q}(y, z) = \prod_{i=1}^{k} (c_i z - b_i y)^{e_i}$$

say, where each e_i is a positive integer,

$$e_1 + \ldots + e_k = nm,$$

and (b_i, c_i) is not a scalar multiple of (b_j, c_j) when $i \neq j$. By the argument used to prove theorems 3.8 and 3.9 there exist unique complex numbers a_i such that

$$C \cap D = \{p_i : 1 \leq i \leq k\}$$

where $p_i = [a_i, b_i, c_i]$, and

$$I_{p_i}(C, D) = e_i.$$

The result follows. $\square$

We can now describe exactly when the intersection multiplicity $I_p(C, D)$ is one.

Proposition 3.22 *Let C and D be projective curves in $\mathbf{P}_2$ and let p be any point in $\mathbf{P}_2$. Then $I_p(C, D) = 1$ if and only if p is a nonsingular point of C and of D and the tangent lines to C and D at p are distinct.*

Remark 3.23 The proof of this proposition can be extended to show that in general

$$I_p(C, D) \geq m_p(C) m_p(D)$$

where $m_p(C)$ and $m_p(D)$ are the multiplicities of C and D at p as defined in exercise 2.8, and that equality holds if and only if C and D have no tangent lines in common.

In order to prove proposition 3.22 we first need to prove the following lemma.

Lemma 3.24 *If $p \in C \cap D$ is a singular point of C then $I_p(C, D) > 1$.*

Proof. We may assume that C and D have no common component, and hence we may choose coordinates such that $p = [0,0,1]$ and the conditions (a)–(c) of theorem 3.18 hold. We wish to show that y^2 divides the resultant $\mathcal{R}_{P,Q}(y,z)$ of the polynomials $P(x,y,z)$ and $Q(x,y,z)$ defining C and D. Since $p \in Sing(C)$ we have

$$\frac{\partial P}{\partial x}(0,0,1) = \frac{\partial P}{\partial y}(0,0,1) = P(0,0,1) = 0.$$

Hence $P(x,y,z)$ is a sum of monomials all of degree at least two in x and y; i.e.

$$P(x,y,z) = a_0(y,z) + a_1(y,z)x + ... + a_n(y,z)x^n$$

where y^2 divides $a_0(y,z)$ and y divides $a_1(y,z)$. Also $Q(0,0,1) = 0$ so

$$Q(x,y,z) = b_0(y,z) + b_1(y,z)x + ... + b_m(y,z)x^m$$

where y divides $b_0(y,z)$. Thus we can write

$$b_0(y,z) = b_{01}yz^{m-1} + y^2c_0(y,z),$$

and

$$b_1(y,z) = b_{10}z^{m-1} + yc_1(y,z)$$

for some homogeneous polynomials $c_0(y,z)$ and $c_1(y,z)$. If $b_{01} = 0$ then the first column of the determinant defining $\mathcal{R}_{P,Q}(y,z)$ is divisible by y^2 and hence y^2 divides $\mathcal{R}_{P,Q}(y,z)$ as required. If $b_{01} \neq 0$ then the first column is divisible by y; if we take out this factor y and subtract b_{10}/b_{01} times the first column from the second column then the second column becomes divisible by y. Hence again y^2 divides $\mathcal{R}_{P,Q}(y,z)$. $\square$

Proof of proposition 3.22. We may assume that p belongs to $C \cap D$ and that C and D have no common component. Thus we may choose coordinates such that the conditions (a)–(c) of theorem 3.18 are satisfied and $p = [0,0,1]$. By corollary 3.10 we may also assume that p is a nonsingular point of C and of D. Let $P(x,y,z)$ and $Q(x,y,z)$ be the polynomials defining C and D. We wish to show that the tangent lines to C and D at p coincide if and only if y^2 divides the resultant $\mathcal{R}_{P,Q}(y,z)$, or equivalently if and only if

$$\frac{\partial \mathcal{R}_{P,Q}}{\partial y}(0,1) = 0$$

since $\mathcal{R}_{P,Q}(y,z)$ is homogeneous and divisible by y.

By (c) of theorem 3.18, the point $[1,0,0]$ does not lie on the tangent line

$$x\frac{\partial P}{\partial x}(0,0,1) + y\frac{\partial P}{\partial y}(0,0,1) + z\frac{\partial P}{\partial z}(0,0,1) = 0$$

to C at $p = [0,0,1]$, so

$$\frac{\partial P}{\partial x}(0,0,1) \neq 0. \tag{3.1}$$

Therefore by the implicit function theorem for complex polynomials (see Appendix B) applied to the polynomial $P(x,y,1)$ in x and y, there is a holomorphic function $\lambda_1 : U \to V$ where U and V are open neighbourhoods of 0 in $\mathbf{C}$ such that

$$\lambda_1(0) = 0$$

and if $x \in V$ and $y \in U$ then

$$P(x,y,1) = 0$$

if and only if

$$x = \lambda_1(y).$$

Moreover

$$P(x,y,1) = (x - \lambda_1(y))l(x,y)$$

where $l(x,y)$ is a polynomial in x whose coefficients are holomorphic functions of y. If we assume as we may that the coefficient $P(1,0,0)$ of x^n in $P(x,y,z)$ is one then

$$l(x,y) = \prod_{i=2}^{n}(x - \lambda_i(y))$$

where $\lambda_1(y),..., \lambda_n(y)$ are the roots of $P(x,y,1)$ regarded as a polynomial in x with y fixed. Similarly if U and V are chosen small enough there is a holomorphic function $\mu_1 : U \to V$ such that

$$\mu_1(0) = 0$$

and we can write

$$Q(x,y,1) = (x - \mu_1(y))m(x,y)$$

where

$$m(x,y) = \prod_{i=2}^{m}(x - \mu_i(y))$$

is a polynomial in x whose coefficients are holomorphic functions of y. Then the tangent lines to C and D at $p = [0,0,1]$ are defined by the equations

$$x = \lambda_1'(0)y$$

and

$$x = \mu_1'(0)y.$$

If $y \in U$ then by lemma 3.6

$$\mathcal{R}_{P,Q}(y,1) = (\mu_1(y) - \lambda_1(y))S(y) \tag{3.2}$$

where

$$S(y) = \prod_{(i,j)\neq(1,1)} (\mu_i(y) - \lambda_j(y)).$$

Note that $S(y)$ is the product of the resultants of the pairs of polynomials $l(x,y)$ and $m(x,y)$, $l(x,y)$ and $x - \mu_1(y)$, and $m(x,y)$ and $x - \lambda_1(y)$. Hence $S(y)$ is a holomorphic function of $y \in U$.

Since $\lambda_1(0) = 0 = \mu_1(0)$ we obtain on differentiating equation 3.2 that

$$\frac{\partial \mathcal{R}_{P,Q}}{\partial y}(0,1) = (\mu_1'(0) - \lambda_1'(0))S(0).$$

It follows from the inequality 3.1 that the polynomial $P(x,0,1)$ does not have repeated roots at 0 so if $i > 1$ then

$$\lambda_i(0) \neq 0 = \mu_1(0)$$

and similarly

$$\mu_i(0) \neq 0 = \lambda_1(0).$$

Moreover if $\lambda_i(0) = \mu_j(0)$ for some $i,j > 1$ then $[0,0,1]$ and

$$[\lambda_i(0),0,1] = [\mu_j(0),0,1]$$

are distinct points of $C \cap D$ both lying on the line $y = 0$, which contradicts condition (b) of theorem 3.18. Hence

$$S(0) \neq 0.$$

Thus

$$\frac{\partial \mathcal{R}_{P,Q}}{\partial y}(0,1) = 0$$

if and only if

$$\lambda_1'(0) = \mu_1'(0),$$

i.e. $I_p(C,D) > 1$ if and only if the tangents to C and D at p coincide. $\square$

Corollary 3.25 *Let C and D be projective curves in $\mathbf{P}_2$ of degrees n and m. Suppose that every $p \in C \cap D$ is a nonsingular point of C and of D, and that the tangent lines to C and D at p are distinct. Then the intersection $C \cap D$ consists of exactly nm points.*

Proof. This follows directly from theorem 3.1 and proposition 3.22. $\square$

Remark 3.26 Even if we allow curves to be defined by polynomials with repeated factors, the conditions of this corollary are never satisfied if either of the polynomials $P(x, y, z)$ and $Q(x, y, z)$ defining C and D have repeated factors. For suppose $P(x, y, z) = A(x, y, z)^2 B(x, y, z)$ where $A(x, y, z)$ is a nonconstant irreducible polynomial. Then the curve defined by $A(x, y, z)$ meets D in at least one point p. It is easy to check that $p \in C \cap D$ and all the partial derivatives of P vanish at p.

We shall finish this section by proving lemmas 3.3–3.7.

Proof of lemma 3.3. Let

$$P(x) = a_0 + a_1 x + \dots + a_n x^n$$

and

$$Q(x) = b_0 + b_1 x + \dots + b_m x^m$$

be polynomials of degrees n and m in x. Then $P(x)$ and $Q(x)$ have a nonconstant common factor $R(x)$ if and only if there exist polynomials $\phi(x)$ and $\psi(x)$ such that

$$P(x) = R(x)\phi(x), Q(x) = R(x)\psi(x).$$

This happens if and only if there exist nonzero polynomials

$$\phi(x) = \alpha_0 + \alpha_1 x + \dots + \alpha_{n-1} x^{n-1}$$

and

$$\psi(x) = \beta_0 + \beta_1 x + \dots + \beta_{m-1} x^{m-1}$$

of degrees at most $n - 1$ and $m - 1$, such that

$$P(x)\psi(x) = Q(x)\phi(x).$$

Equating the coefficients of x^j in this equation for $0 \le j \le mn - 1$, we find

$$a_0 \beta_0 = b_0 \alpha_0$$
$$a_0 \beta_1 + a_1 \beta_0 = b_1 \alpha_0 + b_0 \alpha_1$$
$$\cdot \quad \cdot \quad \cdot$$
$$a_n \beta_{m-1} = b_m \alpha_{n-1}.$$

The existence of a nonzero solution

$$(\alpha_0, \dots, \alpha_{n-1}, \beta_0, \dots, \beta_{m-1})$$

to these equations is equivalent to the vanishing of the determinant which defines $\mathcal{R}_{P,Q}$. $\square$

Proof of lemma 3.4. Let $P(x,y,z)$ and $Q(x,y,z)$ be nonconstant homogeneous polynomials in x,y,z of degrees n and m such that

$$P(1,0,0) \neq 0 \neq Q(1,0,0).$$

We may assume that $P(1,0,0) = 1 = Q(1,0,0)$. Then we can regard P and Q as monic polynomials of degrees n and m in x with coefficients in the ring $C[y,z]$ of polynomials in y and z with complex coefficients. This ring $C[y,z]$ is contained in the field $C(y,z)$ of rational functions of y and z, i.e. functions of the form

$$\frac{f(y,z)}{g(y,z)}$$

where $f(y,z)$ and $g(y,z)$ are polynomials and $g(y,z)$ is not identically zero. Since $C(y,z)$ is a field the proof of lemma 3.3 shows that the resultant $\mathcal{R}_{P,Q}(y,z)$ vanishes identically if and only if $P(x,y,z)$ and $Q(x,y,z)$ have a nonconstant common factor when regarded as polynomials in x with coefficients in $C(y,z)$. It follows from the Gauss lemma (see Appendix A) that this happens if and only if $P(x,y,z)$ and $Q(x,y,z)$ have a nonconstant common factor when regarded as polynomials in x with coefficients in $C[y,z]$, or equivalently as polynomials in x,y,z with coefficients in C. Since any polynomial factor of a homogeneous polynomial is homogeneous (see Appendix A) the result follows. $\square$

Proof of lemma 3.6. If we regard

$$P(x) = (x - \lambda_1)...(x - \lambda_n)$$

and

$$Q(x) = (x - \mu_1)...(x - \mu_m)$$

as homogeneous polynomials in x, $\lambda_1, ..., \lambda_n$ and $\mu_1, ..., \mu_m$ then the proof of lemma 3.7 shows that the resultant $\mathcal{R}_{P,Q}$ is a homogeneous polynomial of degree nm in the variables

$$\lambda_1, ..., \lambda_n, \mu_1, ..., \mu_m.$$

Moreover by lemma 3.3 this polynomial vanishes if $\lambda_i = \mu_j$ for any $1 \leq i \leq n$ and $1 \leq j \leq m$, so by A4 and A6 it is divisible by

$$\prod_{1 \leq i \leq n, 1 \leq j \leq m} (\mu_i - \lambda_j).$$

Since this is also a homogeneous polynomial of degree nm in $\lambda_1, ..., \lambda_n$ and $\mu_1, ..., \mu_m$ it must be a scalar multiple of $\mathcal{R}_{P,Q}$. It is easy to check that if

$$\mu_1 = ... = \mu_m = 0$$

so that $Q(x) = x^m$ then

$$\mathcal{R}_{P,Q} = \prod_{i=1}^{n} (-\lambda_i)^m.$$

Hence we must have

$$\mathcal{R}_{P,Q} = \prod_{i,j} (\mu_i - \lambda_j).$$

It follows immediately that

$$\mathcal{R}_{P,QR} = \mathcal{R}_{P,Q} \mathcal{R}_{P,R}$$

whenever P, Q, R are polynomials in x. Therefore if P, Q, R are polynomials in x, y, z we have

$$\mathcal{R}_{P,QR}(b,c) = \mathcal{R}_{P,Q}(b,c) \mathcal{R}_{P,R}(b,c)$$

for all $b, c \in \mathbf{C}$, and so

$$\mathcal{R}_{P,QR}(y,z) = \mathcal{R}_{P,Q}(y,z) \mathcal{R}_{P,R}(y,z)$$

as required. $\square$

Proof of lemma 3.7. By definition the resultant $\mathcal{R}_{P,Q}(y,z)$ of homogeneous polynomials $P(x,y,z)$ and $Q(x,y,z)$ of degrees n and m is the determinant of an $n + m$ by $n + m$ matrix whose ijth entry $r_{ij}(y,z)$ is a homogeneous polynomial in y and z of degree d_{ij} given by

$$d_{ij} = \begin{cases} n + i - j & \text{if } 1 \leq i \leq m \\ i - j & \text{if } m + 1 \leq i \leq n + m. \end{cases}$$

Then $\mathcal{R}_{P,Q}(y,z)$ is a sum of terms of the form

$$\pm \prod_{i=1}^{n+m} r_{i\sigma(i)}(y,z)$$

where σ is a permutation of $\{1, ..., n + m\}$. Each such term is a homogeneous polynomial of degree

$$\begin{aligned}
\sum_{i=1}^{n+m} d_{i\sigma(i)} &= \sum_{i=1}^{m} (n + i - \sigma(i)) + \sum_{i=m+1}^{m+n} (i - \sigma(i)) \\
&= nm + \sum_{i=1}^{m+n} i - \sum_{i=1}^{m+n} \sigma(i) \\
&= nm.
\end{aligned}$$

Therefore $\mathcal{R}_{P,Q}(y,z)$ is a homogeneous polynomial of degree nm in y and z.
$\square$

3.2 Points of inflection and cubic curves.

The concept of a point of inflection on a curve C is a generalisation of the usual definition of a point of inflection on the graph of a function (that is, a point at which the second derivative vanishes). We shall see that every nonsingular projective curve of degree greater than two has at least one and at most finitely many points of inflection. As a corollary we shall show that every nonsingular projective cubic curve can be put into the form

$$y^2 z = x(x - z)(x - \lambda z)$$

for some $\lambda \in \mathbf{C} - \{0, 1\}$. We shall also begin our investigation of the natural abelian group structure on a nonsingular cubic curve, and its relationship with the points of inflection on the curve.

Definition 3.27 *Let $P(x, y, z)$ be a homogeneous polynomial of degree d. The* Hessian $\mathcal{H}_P$ *of P is the polynomial defined by*

$$\mathcal{H}_P(x, y, z) = \det \begin{pmatrix} P_{xx} & P_{xy} & P_{xz} \\ P_{yx} & P_{yy} & P_{yz} \\ P_{zx} & P_{zy} & P_{zz} \end{pmatrix}$$

where

$$P_x = \frac{\partial P}{\partial x},$$

$$P_{xx} = \frac{\partial^2 P}{\partial x^2},$$

$$P_{xy} = \frac{\partial^2 P}{\partial x \partial y} \quad \text{etc.}$$

Remark 3.28 Note that the second partial derivatives of P are homogeneous of degree $d - 2$ in x, y, z so $\mathcal{H}_P$ is a homogeneous polynomial of degree $3(d-2)$ in x, y, z.

Definition 3.29 *A nonsingular point $[a, b, c]$ of the projective curve C in $\mathbf{P}_2$ defined by $P(x, y, z)$ is called a* point of inflection *(or flex) of C if*

$$\mathcal{H}_P(a, b, c) = 0.$$

In order to see how this definition relates to the usual definition of an inflection point on a graph, we need the next lemma.

Lemma 3.30 *If $P(x, y, z)$ is a homogeneous polynomial of degree $d > 1$ then*

$$z^2 \mathcal{H}_P(x, y, z) = (d - 1)^2 \det \begin{pmatrix} P_{xx} & P_{xy} & P_x \\ P_{yx} & P_{yy} & P_y \\ P_x & P_y & dP/(d-1) \end{pmatrix}.$$

Proof. Euler's relation (lemma 2.32) tells us that

$$dP(x, y, z) = xP_x(x, y, z) + yP_y(x, y, z) + zP_z(x, y, z).$$

Since the first partial derivatives of P are homogeneous of degree $d - 1$ we can apply Euler's relation to them to get

$$\begin{array}{rcl}
(d-1)P_x &=& xP_{xx} + yP_{yx} + zP_{zx}, \\
(d-1)P_y &=& xP_{xy} + yP_{yy} + zP_{zy}, \\
(d-1)P_z &=& xP_{xz} + yP_{yz} + zP_{zz}.
\end{array}$$

Thus multiplying the first row of the determinant defining $\mathcal{H}_P$ by x and the second row by y and adding to the third row multiplied by z we get

$$z\mathcal{H}_P(x, y, z) = (d-1) \det \begin{pmatrix} P_{xx} & P_{xy} & P_{xz} \\ P_{yx} & P_{yy} & P_{yz} \\ P_x & P_y & P_z \end{pmatrix}.$$

Applying the same procedure to the columns of this new determinant and using the fact that the second partial derivatives of P are symmetric we get the required result. $\square$

Remark 3.31 Where the partial derivative $\frac{\partial P}{\partial y}$ is nonzero, the equation

$$P(x, y, 1) = 0$$

locally defines y as a holomorphic function of x (by the implicit function theorem in Appendix B). Differentiating this equation twice with respect to x we obtain first

$$\frac{\partial P}{\partial x} + \frac{dy}{dx}\frac{\partial P}{\partial y} = 0,$$

i.e.

$$\frac{dy}{dx} = -\frac{\partial P}{\partial x} \Big/ \frac{\partial P}{\partial y},$$

and then

$$\frac{\partial^2 P}{\partial x^2} + \left(\frac{dy}{dx}\right)^2 \frac{\partial^2 P}{\partial y^2} + 2\frac{dy}{dx}\frac{\partial^2 P}{\partial x \partial y} + \frac{d^2 y}{dx^2}\frac{\partial P}{\partial y} = 0,$$

i.e.

$$\frac{d^2 y}{dx^2} = \frac{2P_x P_y P_{xy} - (P_y)^2 P_{xx} - (P_x)^2 P_{yy}}{(P_y)^3}$$

$$= (P_y)^{-3} \det \begin{pmatrix} P_{xx} & P_{xy} & P_x \\ P_{yx} & P_{yy} & P_y \\ P_x & P_y & 0 \end{pmatrix}.$$

It therefore follows immediately from lemma 3.30 that if $d > 1$ then

$$\frac{d^2y}{dx^2} = \frac{\mathcal{H}_P(x,y,1)}{(d-1)^2(P_y)^3}.$$

In particular if $P(a,b,1) = 0 \neq P_y(a,b,1)$ then $[a,b,1]$ is a point of inflection of C if and only if $\frac{d^2y}{dx^2}$ vanishes at a when y is regarded as a function of x implicitly defined by the equation $P(x,y,1) = 0$. Thus the definition of a point of inflection on a curve does correspond in a reasonable way to the definition of a point of inflection on a graph.

Lemma 3.32 *Let*

$$C = \{[x,y,z] \in \mathbf{P}_2 \ : \ P(x,y,z) = 0\}$$

be an irreducible projective curve of degree d. Then every point of C is a point of inflection if and only if $d = 1$.

Proof. Suppose that every point of C is a point of inflection. By applying a suitable projective transformation we may assume that

$$P(0,0,1) = 0 \neq \frac{\partial P}{\partial y}(0,0,1).$$

Then the implicit function theorem (see Appendix A) applied to the polynomial $P(x,y,1)$ in x and y tells us that there is a holomorphic function $g : U \to V$, where U and V are open neighbourhoods of 0 in $\mathbf{C}$, such that $g(0) = 0$ and if $x \in U$ and $y \in V$ then

$$P(x,y,1) = 0$$

if and only if

$$y = g(x).$$

We may assume that U is connected and that $P_y(x,y,1) \neq 0$ when $x \in U$ and $y \in V$. Since every point of C is an inflection point, by remark 3.31 we have

$$g''(x) = 0$$

for every $x \in U$. Because $g(0) = 0$ this means that there is some $\lambda \in \mathbf{C}$ such that

$$g(x) = \lambda x$$

for all $x \in U$, and hence that the polynomial $P(x, \lambda x, 1)$ in x vanishes identically. Since $P(x,y,z)$ is homogeneous we find by equating the coefficients of x^j in $P(x, \lambda x, 1)$ to zero that $P(x,y,z)$ is divisible by $y - \lambda x$. But C is irreducible so $P(x,y,z)$ must be a scalar multiple of $y - \lambda x$. Thus $P(x,y,z)$ has degree $d = 1$ as required.

The converse is trivial, so the proof is complete. $\square$

Proposition 3.33 *Let C be a nonsingular projective curve in $\mathbf{P}_2$ of degree d.*
(i) If $d \geq 2$ then C has at most $3d(d-2)$ points of inflection.
(ii) If $d \geq 3$ then C has at least one point of inflection.

Proof. By remark 3.28 $\mathcal{H}_P$ is homogeneous of degree $3(d-2)$, so provided that it is not constant (when $d > 2$ this means not identically zero) it defines a projective curve in $\mathbf{P}_2$ in the generalised sense of remarks 2.3 and 2.25. Since the whole of §3.1 applies to curves in this generalised sense the result follows from the forms 3.8 and 3.9 of Bézout's theorem once we have shown that P and $\mathcal{H}_P$ have no nonconstant common factor if $d > 1$. Since a nonsingular curve is irreducible (by corollary 3.10(i)) if P and $\mathcal{H}_P$ have a nonconstant common factor then P divides $\mathcal{H}_P$, so every point of C is a point of inflection. The result is now a consequence of lemma 3.32. □

Corollary 3.34 *Let C be a nonsingular cubic curve in $\mathbf{P}_2$. Then C is equivalent under a projective transformation to the curve defined by*

$$y^2 z = x(x-z)(x-\lambda z)$$

for some $\lambda \in \mathbf{C} - \{0,1\}$.

Proof. By proposition 3.33 C has a point of inflection. By applying a suitable projective transformation we may assume that $[0,1,0]$ is a point of inflection of C and the tangent line to C at $[0,1,0]$ is the line $z = 0$. Then C is defined by a homogeneous polynomial $P(x,y,z)$ of degree three such that

$$P(0,1,0) = 0 = \frac{\partial P}{\partial x}(0,1,0) = \frac{\partial P}{\partial y}(0,1,0) = \mathcal{H}_P(0,1,0).$$

Also

$$\frac{\partial P}{\partial z}(0,1,0) \neq 0$$

because C is nonsingular. Applying lemma 3.30 with the roles of y and z reversed, we get

$$y^2 \mathcal{H}_P(x,y,z) = 4 \det \begin{pmatrix} P_{xx} & P_x & P_{xz} \\ P_x & \frac{3}{2}P & P_z \\ P_{zx} & P_z & P_{zz} \end{pmatrix}$$

so

$$0 = \mathcal{H}_P(0,1,0) = 4 \det \begin{pmatrix} P_{xx} & 0 & P_{xz} \\ 0 & 0 & P_z \\ P_{zx} & P_z & P_{zz} \end{pmatrix} = -4(P_z)^2 P_{xx},$$

where the partial derivatives are evaluated at $(0, 1, 0)$. Thus

$$P_{xx}(0, 1, 0) = 0$$

and hence

$$P(x, y, z) = yz(\alpha x + \beta y + \gamma z) + \phi(x, z)$$

where $\phi(x, z)$ is homogeneous of degree three in x and z and

$$\beta = \frac{\partial P}{\partial z}(0, 1, 0) \neq 0.$$

After the projective transformation given by

$$[x, y, z] \mapsto [x, y + \frac{\alpha x + \gamma z}{2\beta}, z],$$

the curve C is defined by the equation

$$\beta y^2 z + \psi(x, z) = 0$$

where $\psi(x, z)$ is homogeneous of degree three in x and z, and hence is a product of three linear factors. Since C is nonsingular it is irreducible (by corollary 3.10(i)), and hence $\psi(x, z)$ is not divisible by z so the coefficient of x^3 in $\psi(x, z)$ is not zero. Therefore after a suitable diagonal projective transformation C is defined by the equation

$$y^2 z = (x - az)(x - bz)(x - cz)$$

for some $a, b, c \in \mathbf{C}$. Since a, b, c are distinct (otherwise C would be singular) we can apply the projective transformation

$$[x, y, z] \mapsto [\frac{x - az}{b - a}, \eta y, z]$$

where $\eta \in \mathbf{C}$ satisfies

$$\eta^2 = (b - a)^{-3}$$

to put C into the form

$$y^2 z = x(x - z)(x - \lambda z)$$

for some $\lambda \in \mathbf{C} - \{0, 1\}$. $\square$

Remark 3.35 Note that the proof of corollary 3.34 actually shows that if p is any point of inflection on a nonsingular cubic curve C then there is a projective transformation taking p to $[0, 1, 0]$ and taking C to a curve defined by

$$y^2 z = x(x - z)(x - \lambda z)$$

for some $\lambda \in \mathbf{C} - \{0, 1\}$.

Example 3.36 As an application of this last result and the corollary 3.25 to Bézout's theorem we can show that a nonsingular cubic curve C in $\mathbf{P}_2$ has *exactly* nine points of inflection. (Thus the bound given in proposition 3.33(i) is always attained in the case of a nonsingular cubic.) To prove this we suppose that C is defined by $P(x,y,z)$ and we let D be the projective curve in $\mathbf{P}_2$ defined by the Hessian $\mathcal{H}_P(x,y,z)$. We know from remark 3.28 that $\mathcal{H}_P(x,y,z)$ is homogeneous of degree 3; however it may have repeated factors so as in the proof of proposition 3.33 D must be regarded as a curve in the generalised sense of remarks 2.3 and 2.25. Since the whole of §3.1 applies to curves in this generalised sense, it suffices to show that the conditions of corollary 3.25 are satisfied. That is, we have to show that if $p \in C \cap D$, or equivalently if p is a point of inflection of C, then p is a nonsingular point of C and of D and the tangent lines to C and D at p are distinct.

We know from the last remark that by applying a suitable projective transformation we can assume that

$$p = [0,1,0]$$

and

$$P(x,y,z) = y^2 z - x(x-z)(x-\lambda z)$$

for some $\lambda \in \mathbf{C} - \{0,1\}$. We find that

$$\frac{\partial P}{\partial x}(0,1,0) = 0 = \frac{\partial P}{\partial y}(0,1,0),$$

$$\frac{\partial P}{\partial z}(0,1,0) = 1,$$

$$\frac{\partial \mathcal{H}_P}{\partial x}(0,1,0) = 24,$$

$$\frac{\partial \mathcal{H}_P}{\partial y}(0,1,0) = 0,$$

and

$$\frac{\partial \mathcal{H}_P}{\partial z}(0,1,0) = -8(\lambda + 1).$$

Thus the conditions of corollary 3.25 are satisfied.

As another useful result about points of inflection on cubics we prove

Lemma 3.37 *A line L in $\mathbf{P}_2$ meets a nonsingular cubic C either*
(a) in three distinct points p, q, r each with intersection multiplicity one (i.e. L is not the tangent line to C at p, q or r); or

(b) in two points, p with intersection multiplicity one and q with intersection multiplicity two (i.e. L is the tangent line to C at q but not at p and q is not a point of inflection on C); or
(c) in one point p with intersection multiplicity three (i.e. L is the tangent line to C at p and p is a point of inflection on C).

Proof. This can be deduced from the strong form 3.1 of Bézout's theorem by checking that the definition of intersection multiplicities given in the statement of the lemma coincides with the definition given in theorem 3.18. However it is hardly longer and perhaps more illuminating to give a direct proof, so that is what will be done here.

Since C is irreducible (by corollary 3.10(i)) it does not contain L, so we may assume that L is the line defined by

$$y = 0$$

and the point $[1, 0, 0]$ does not belong to C. Let

$$C = \{[x, y, z] \in \mathbf{P}_2 \; : \; P(x, y, z) = 0\}.$$

Then by lemma 2.8 we can factorise $P(x, 0, z)$ as

$$P(x, 0, z) = \mu(x - \lambda_1 z)(x - \lambda_2 z)(x - \lambda_3 z)$$

for some $\lambda_1, \lambda_2, \lambda_3 \in \mathbf{C}$ and $\mu \in \mathbf{C} - \{0\}$, so

$$\begin{aligned} C \cap L &= \{[x, 0, z] \in \mathbf{P}_2 \; : \; P(x, 0, z) = 0\} \\ &= \{[\lambda_i, 0, 1] \; : \; 1 \le i \le 3\}. \end{aligned}$$

The tangent line to C at $[\lambda_i.0, 1]$ is defined by

$$x \frac{\partial P}{\partial x}(\lambda_i, 0, 1) + y \frac{\partial P}{\partial y}(\lambda_i, 0, 1) + z \frac{\partial P}{\partial z}(\lambda_i, 0, 1) = 0.$$

By Euler's relation (lemma 2.32) this line is L if and only if

$$\frac{\partial P}{\partial x}(\lambda_i, 0, 1) = 0,$$

or equivalently if and only if λ_i is a repeated root of the polynomial

$$P(x, 0, 1) = \mu(x - \lambda_1)(x - \lambda_2)(x - \lambda_3).$$

If so then by lemma 3.30

$$\mathcal{H}_P(\lambda_i, 0, 1) = 4 \det \begin{pmatrix} P_{xx} & P_{xy} & 0 \\ P_{yx} & P_{yy} & P_y \\ 0 & P_y & 0 \end{pmatrix}$$

$$= -4(P_y)^2 P_{xx}$$

evaluated at $(0,1,0)$, and

$$\frac{\partial P}{\partial y}(\lambda_i, 0, 1) \neq 0$$

since C is nonsingular; so $[\lambda_i, 0, 1]$ is a point of inflection if and only if

$$\frac{\partial^2 P}{\partial x^2}(\lambda_i, 0, 1) = 0,$$

or equivalently if and only if λ_i is a root of multiplicity three in the polynomial $P(x, 0, 1)$. The result follows. $\square$

We end this chapter with a very pretty result which says that nonsingular projective curves in $\mathbf{P}_2$ have natural abelian group structures. We will not quite be able to complete the proof here (though see exercises 3.13 and 3.14) but this will be done later (see 6.21 and 6.39).

Theorem 3.38 *Given any nonsingular projective cubic C in $\mathbf{P}_2$ and a point of inflection p_0 on C there is a unique additive group structure on C such that p_0 is the zero element and three points of C add up to zero if and only if they are the three points of intersection of C with some line in $\mathbf{P}_2$ (allowing for multiplicities).*

Proof. To check uniqueness, note first that additive inverses are uniquely determined since $-p_0 = p_0$ and if $p \neq p_0$ then $-p$ is the third point of intersection of C with the line in $\mathbf{P}_2$ through p and p_0. Also if p, q are any points of C then $p + q = -r$ where r is the third point of intersection of C with the line in $\mathbf{P}_2$ through p and q (if $p \neq q$) or the tangent line to C at p (if $p = q$). Thus the additive group structure is uniquely determined.

It remains to show there is an additive group structure with p_0 as zero defined in this way. Commutativity comes straight from the definition of $p + q$. For any $p \in C$ such that $p \neq p_0$ we have $p + p_0 = -r$ where r is the third point of intersection of C with the line in $\mathbf{P}_2$ through p and p_0. This point r is not p_0 since p_0 is a point of inflection, so $-r$ is the third point of intersection of C with the line in $\mathbf{P}_2$ through r and p_0, which is of course p. Thus $p + p_0 = p$ if $p \neq p_0$, and $p_0 + p_0 = p_0$ since p_0 is a point of inflection (so that its tangent line meets C with multiplicity three at p_0). The proof that $p + (-p) = p_0$ for every $p \in C$ comes equally easily from the definitions of addition and inverses, so it only remains to prove associativity.

We cannot prove associativity at this point. There are many different methods of proof, two of which will be given later in this book (see remark 6.21 and theorem 6.39). One more is left to the exercises (see exercises 3.13 and 3.14). $\square$

3.3   Exercises

3.1. Let C and D be projective curves in $\mathbf{P}_2$ with no common component. Show that

$$Sing(C \cup D) = Sing(C) \cup Sing(D) \cup (C \cap D).$$

Use corollary 3.10 to deduce that any projective curve in $\mathbf{P}_2$ defined by a polynomial with no repeated factors has at most finitely many singular points.

3.2. Use Bézout's theorem (theorem 3.1 and lemma 3.25) to show that if a projective curve C in $\mathbf{P}_2$ of degree d has strictly more than $d/2$ singular points all lying on a line L then L is a component of C.

3.3. Show that given any five points in $\mathbf{P}_2$ there is at least one conic containing them. Deduce that a projective curve C of degree four in $\mathbf{P}_2$ with four singular points is reducible. [Hint: show that any conic containing the four singular points and another point of C must have a component in common with C].

3.4. Show that a polynomial $P(x)$ in one variable with complex coefficients has a repeated factor if and only if the resultant

$$\mathcal{R}_{P,P'}$$

is zero. Show also that if

$$P(x) = \Pi_{i=1}^{n}(x - \alpha_i)$$

then

$$
\begin{aligned}
\mathcal{R}_{P,P'} &= \Pi_{i=1}^{n} \mathcal{R}_{x-\alpha_i,P'} \\
&= \Pi_{i=1}^{n} P'(\alpha_i) \\
&= (-1)^{\frac{1}{2}n(n-1)} \left(\Pi_{i<j}(\alpha_i - \alpha_j)\right)^2
\end{aligned}
$$

3.5. Prove the following converse to Pascal's theorem: if the intersections of the opposite sides of a hexagon lie on a straight line then the vertices lie on a conic. [This is to be interpreted as follows. Let $p_1, \ldots, p_6$ be six distinct points of $\mathbf{P}_2$, no three of which lie on a line. If $i \neq j$ let L_{ij} be the line through p_i and p_j. It is required to show that if the three points of intersection of the pairs of lines L_{12} and L_{45}, L_{23} and L_{56}, L_{34} and L_{61} all lie on a line then the points $p_1, \ldots, p_6$ lie on a conic. You may wish to consider cubic curves which are unions of lines in $\mathbf{P}_2$].

3.6. Prove Pappus' theorem: if L and M are two projective lines in $\mathbf{P}_2$ and p_1, p_2, p_3 lie on $L - L \cap M$ and q_1, q_2, q_3 lie on $M - L \cap M$ then if L_{ij} is the line joining p_i and q_j the three points of intersection of the pairs of lines L_{ij} and L_{ji} are collinear. (This is almost the same as Pascal's theorem and can be proved in almost the same way. The only difference is that the conic is now the union of the lines L and M and hence is reducible).

3.7. If in Pascal's theorem we let some vertices of the hexagon coincide (the corresponding side of the hexagon becoming a tangent to the conic) we get new theorems. State (and sketch in $\mathbf{R}^2$) what happens if

(a). $p_1 = p_2, \quad p_3 = p_4, \quad p_5 = p_6$, where $p_1, \dots, p_6$ are the vertices of the hexagon;

(b). $p_1 = p_2$ and all the four other vertices are distinct.

From (b) deduce a rule for constructing a tangent to a given conic in $\mathbf{R}^2$ at a given point using only a straight edge.

3.8. Let C be a projective curve in $\mathbf{P}_2$ defined by a homogeneous polynomial $P(x, y, z)$ and let α be a linear transformation of $\mathbf{C}^3$. Let Q be the homogeneous polynomial $Q = P \circ \alpha^{-1}$ which defines the image of C under the projective transformation given by α. Show that the matrix of second derivatives of Q at a point of $\mathbf{P}_2$ represented by $v \in \mathbf{C}^3 - \{0\}$ is given by pre- and post-multiplying the matrix of second derivatives of P at the point represented by $\alpha^{-1}(v)$ by the matrix of the linear transformation α^{-1} and its transpose, and hence that

$$\mathcal{H}_P \circ \alpha^{-1} = (det\ \alpha)^2 \mathcal{H}_Q.$$

Deduce that the definition of an inflection point is invariant under projective transformations.

3.9. Let C be a projective cubic curve with a singularity at the point $[0, 0, 1]$. Show that the equation of C is of the form

$$\text{(quadratic in } x \text{ and } y)z = \text{cubic in } x \text{ and } y.$$

Show that by a suitable change of coordinates the equation can be put into one of the forms

$$y^2 z = \text{cubic in } x \text{ and } y$$

or

$$xyz = \text{cubic in } x \text{ and } y$$

and hence by a suitable substitution $z \mapsto \lambda x + \mu y + \nu z$ into one of the forms

$$y^2 z = (x + by)^3$$

or

$$xyz = (x+y)^3$$

for some complex number b. By making one more substitution in each case deduce that any irreducible cubic curve in $\mathbf{P}_2$ is equivalent under a projective transformation to one of the following:

$$
\begin{aligned}
y^2 z &= x^3, \\
y^2 z &= x^2(x+z)
\end{aligned}
$$

or

$$y^2 z = x(x-z)(x-\lambda z)$$

for some $\lambda \in \mathbf{C} - \{0\}$ (the last case occurring if the curve is nonsingular).

3.10. What are the singular points of the cubic curves listed in exercise 3.9? What are their points of inflection? Deduce that every irreducible projective cubic curve has a nonsingular point of inflection.

3.11. Show that if p is a point of inflection on a nonsingular cubic curve C in the projective plane then there is a projective transformation taking p to the point $[0,1,0]$ and taking C to a curve of the form

$$y^2 z = 4x^3 - g_2 xz^2 - g_3 z^3$$

where $(g_2)^3 - 27(g_3)^2 \neq 0$.

N.B. This form of the equation is of historical importance and will appear again later in the book, but it is not usually any more convenient than the one in corollary 3.34.

3.12. By comparing the formula for $\mathcal{H}_P$ in remark 3.31 with the corresponding formula for $\mathcal{H}_{PQ}$, or otherwise, show that if C and D are projective curves in $\mathbf{P}_2$ and $p \in C - D$ then p is a point of inflection for the curve C if and only if p is a point of inflection for the curve $C \cup D$.

3.13. Let C, D, E be projective cubic curves in $\mathbf{P}_2$ defined by homogeneous polynomials $P(x,y,z), Q(x,y,z), R(x,y,z)$. Suppose that C and D meet in exactly nine points $p_1, \ldots, p_9$. Show that no line in $\mathbf{P}_2$ contains four of these points and no conic contains seven of them. Show that there is a unique conic Q containing $p_1, \ldots, p_5$.

Now suppose that E contains $p_1, \ldots, p_8$ and that $R(x,y,z)$ is not a linear combination of $P(x,y,z)$ and $Q(x,y,z)$. Show that given distinct points q and r in $\mathbf{P}_2$ there is a curve $C(\lambda, \mu, \nu)$, defined by

$$\lambda P(x,y,z) + \mu Q(x,y,z) + \nu R(x,y,z) = 0$$

with $\lambda, \mu, \nu \in \mathbf{C}$, which passes through $p_1, \ldots, p_8, q$ and r.

If p_8 lies in the line L through p_6 and p_7 choose $q \in L$ and $r \notin L \cup Q$ and obtain a contradiction by showing that $C(\lambda, \mu, \nu) = L \cup Q$. Deduce by symmetry that no three of $p_1, \ldots, p_8$ lie on a line. If $p_8 \notin Q$ obtain a contradiction by choosing q and r in L. Deduce by symmetry that $p_6, p_7, p_8 \in Q$. Conclude that the original hypotheses on E were inconsistent.

Deduce that if E is any projective cubic curve containing $p_1, \ldots, p_8$ then E contains p_9.

3.14. Let C be a nonsingular projective cubic curve in $\mathbf{P}_2$ and let p_0 be a point of inflection on C. Carry out the following argument to complete the proof of theorem 3.38 that there is an additive group structure on C with p_0 as zero such that three points of C add up to zero if and only if they are the points of intersection of C with some line in $\mathbf{P}_2$.

Let p, q, r be any points of C. Let L_1 be the line in $\mathbf{P}_2$ which meets C in the points $p, q, -(p+q)$ counted with multiplicity. Similarly let L_2, L_3, M_1, M_2 and M_3 be the lines in $\mathbf{P}_2$ which meet C in the points

$$
\begin{array}{lll}
p_0, & p+q, & -(p+q), \\
r, & p+q, & -((p+q)+r), \\
q, & r, & -(q+r), \\
p_0, & q+r, & -(q+r), \\
p, & q+r, & -(p+(q+r))
\end{array}
$$

respectively (see figure 3.2). Let D and E be the reducible cubic curves

$$D = L_1 \cup M_2 \cup L_3$$

and

$$E = M_1 \cup L_2 \cup M_3.$$

Show that D meets C in the points

$$p_0, p, q, r, p+q, q+r, -(p+q), -(q+r), -((p+q)+r)$$

whereas E meets C in the points

$$p_0, p, q, r, p+q, q+r, -(p+q), -(q+r), -(p+(q+r)).$$

Now use exercise 3.13 to deduce that

$$(p+q)+r = p+(q+r).$$

[Strictly speaking this argument only works when the points

$$p_0, p, q, r, p+q, q+r, -(p+q), -(q+r), -((p+q)+r)$$

are distinct. It can be extended to the general case either by tightening up the argument of exercise 3.13 to allow for intersection multiplicities greater than

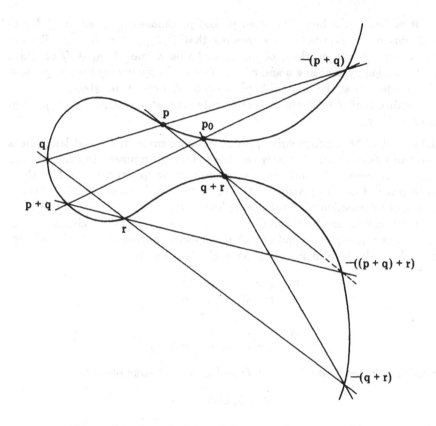

Figure 3.2: Associativity of the group law on a cubic

one, or by a continuity argument using the fact that we can find sequences $p_n \to p, q_n \to q, r_n \to r$ such that the points

$$p_0, p_n, q_n, r_n, p_n + q_n, q_n + r_n, -(p_n + q_n), -(q_n + r_n), -((p_n + q_n) + r_n)$$

are distinct for all $n \geq 1$].

3.15. Let C be the nonsingular cubic curve defined by

$$x^3 + y^3 + z^3 + \lambda xyz = 0$$

where $\lambda^3 + 27 \neq 0$. Show that the points of inflection of C are the points of intersection of C with a different curve of the same form, and deduce that they satisfy

$$x^3 + y^3 + z^3 = 0 = xyz$$

and thus

$$0 = (x + y + z)(x + \varepsilon y + \varepsilon^2 z)(x + \varepsilon^2 y + \varepsilon z)$$

$$= (x + \varepsilon y + \varepsilon z)(x + \varepsilon^2 y + z)(x + y + \varepsilon^2 z)$$
$$= (x + \varepsilon^2 y + \varepsilon^2 z)(x + \varepsilon y + z)(x + y + \varepsilon z)$$

where $\varepsilon = e^{2\pi i/3}$. Hence or otherwise show that C has exactly nine points of inflection (which are independent of λ) and that the line through any two of them meets C again at a third point of inflection.

(N.B. In fact every nonsingular cubic curve in the projective plane is equivalent under a projective transformation to one of the form

$$x^3 + y^3 + z^3 + \lambda xyz = 0).$$

3.16. Use remark 3.35 to show that if p is a point of inflection of a nonsingular cubic curve C in $\mathbf{P}_2$ then there are exactly four tangent lines to C which pass through P.

3.17. Let C be a singular irreducible projective cubic curve in $\mathbf{P}_2$ and let p_0 be a nonsingular point of C which is a point of inflection (cf. exercise 3.10). Modify the proof of theorem 3.38 and exercise 3.14 to show that there is a unique additive group structure on the set of nonsingular points of C with p_0 as the zero element such that three points add up to zero if and only if they are the points of intersection of C with some line in $\mathbf{P}_2$. [Note that a tangent line to C at a nonsingular point or a line through two distinct nonsingular points of C cannot meet C at a singular point by Bézout's theorem (theorem 3.1) and lemma 3.25].

3.18. Let C be the nonsingular cubic defined by

$$y^2 z = x(x - z)(x - \lambda z)$$

for some $\lambda \in \{0, 1\}$ and let $p_0 = [0, 1, 0]$. Show that the additive group structure on C defined as in theorem 3.38 is given by

$$[x_1, y_1, 1] + [x_2, y_2, 1] = \begin{cases} [0, 1, 0] & \text{if } x_1 = x_2 \text{ but } y_1 \neq y_2, \\ [x_3, y_3, 1] & \text{otherwise,} \end{cases}$$

where if $x_1 \neq x_2$ then

$$x_3 = \left(\frac{y_1 - y_2}{x_1 - x_2} \right)^2 + 1 + \lambda - x_1 - x_2$$

and

$$y_3 = \left(\frac{y_1 - y_2}{x_1 - x_2} \right) x_3 + \left(\frac{x_1 y_2 - y_1 x_2}{x_1 - x_2} \right).$$

Find formulas for x_3 and y_3 in the case $x_1 = x_2, y_1 = y_2$.

3.19. Suppose that the polynomial $P(x, y, z)$ defining a nonsingular projective cubic C in $\mathbf{P}_2$ has rational coefficients. Suppose also that p_0 is a point of inflection on C with rational coefficients. Show that if $p, q \in C$ have rational coefficients then so do $-p$ and $p + q$ where addition is defined as in the proof of theorem 3.38. Deduce that

$$C(\mathbf{Q}) = \{[a, b, c] \in C \ : \ a, b, c \in \mathbf{Q}\}$$

is an abelian group.

Chapter 4

Topological properties

As a subset of the projective plane $\mathbf{P}_2$ a complex projective curve

$$C = \{[x, y, z] \in \mathbf{P}_2 \; : \; P(x, y, z) = 0\}$$

has a natural topology, which means that it makes sense to talk about concepts such as continuous functions on C (see §§2.2 and 2.3). In this chapter we shall investigate nonsingular complex projective curves from the topological point of view.

In fact a nonsingular projective curve in $\mathbf{P}_2$ is topologically a sphere with g handles (see figure 4.1 for a picture when $g = 3$). This number g is called the *genus* of the curve. We shall see that it is related to the degree d of the curve by the *degree-genus formula*:

$$g = \frac{1}{2}(d - 1)(d - 2).$$

It is also possible to describe the topology of *singular* projective curves in $\mathbf{P}_2$, although as might be expected the description is more complicated. It is enough to consider *irreducible* curves since any projective curve in $\mathbf{P}_2$ is the union of finitely many irreducible curves (see 2.26) meeting at finitely

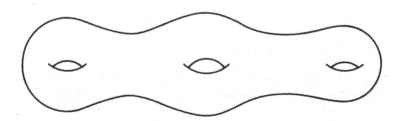

Figure 4.1: A sphere with three handles

85

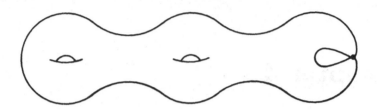

Figure 4.2: A singular curve

many points (by theorem 3.9). It turns out that an irreducible projective curve is the result of making a finite number of identifications of points on a sphere with g handles (see figure 4.2). More precisely, if C is an irreducible projective curve of degree d in $\mathbf{P}_2$ with singular points $p_1, \ldots, p_r$ then there is a sphere with g handles $\tilde{C}$ and a continuous surjection

$$\pi : \tilde{C} \to C$$

which restricts to a homeomorphism

$$\pi : \tilde{C} - \pi^{-1}\{p_1, \ldots, p_r\} \to C - \{p_1, \ldots, p_r\},$$

and $\pi^{-1}(p_i)$ is a finite set of points for each $i \in \{1, \ldots, r\}$. The number g is again called the genus of C. The number of points in $\pi^{-1}(p_i)$ depends on the type of singularity of C at p_i. For example if p_i is an ordinary double point then $\pi^{-1}(p_i)$ consists of two points; more generally if p_i is an ordinary singular point of any multiplicity $m \geq 2$ then $\pi^{-1}(p_i)$ consists of precisely m points. On the other hand if C is the cuspidal cubic curve defined by $y^2 z = x^3$ and p is its only singular point $(0,0)$ then $\pi^{-1}(p)$ consists of just one point and $\pi : \tilde{C} \to C$ is in fact a homeomorphism. The degree-genus formula can be generalised to apply to C; to each singular point p_i there can be assigned a positive integer $\delta(p_i)$ such that the following formula

$$g = \frac{1}{2}(d-1)(d-2) - \sum_{j=1}^{r} \delta(p_j)$$

(called *Noether's formula*) is true.

The proof of these facts about singular curves will be left to Chapter 7. The aim of this chapter is to prove the degree-genus formula for nonsingular projective curves.

There are several different ways to prove the degree-genus formula. In §4.1 we shall give a brief description of two methods. The first is intuitively appealing but it lies beyond the scope of this book to carry it through in detail. The second is essentially the method we shall use in §4.2 and §4.3 to give a detailed proof of the degree-genus formula. In §4.1 the rough idea of this proof is explained and illustrated with an example.

4.1 The degree-genus formula

Our aim is to prove that a nonsingular complex projective curve of degree d in $\mathbf{P}_2$ is topologically a sphere with g handles where the genus g satisfies the *degree-genus formula*

$$g = \frac{1}{2}(d-1)(d-2).$$

In this section we shall describe two methods of proof without details.

4.1.1 The first method of proof

This method is made up of three steps.

The first step is to consider a (singular) complex projective curve C_0 which is a union of d projective lines in $\mathbf{P}_2$, in "general position" in the sense that no point of $\mathbf{P}_2$ lies on more than two of the lines. Thus there are exactly $\frac{1}{2}d(d-1)$ points of intersection of these lines, and these are the singular points of C_0.

Lemma 4.1 *A complex projective line L in $\mathbf{P}_2$ is homeomorphic to the two-dimensional unit sphere*

$$S^2 = \{(u,v,w) \in \mathbf{R}^3 \ : \ u^2 + v^2 + w^2 = 1\}.$$

Proof. We use stereographic projection (cf. §1.2.2). By applying a projective transformation (which is a homeomorphism by lemma 2.20) we may assume that L is the line defined by $z = 0$. Now define

$$\phi : S^2 \to L$$

by

$$\phi(u,v,w) = [u+iv, 1-w, 0].$$

It is not difficult to check using the definition of homogeneous coordinates (definition 2.15) that ϕ is a bijection with inverse given by

$$\phi^{-1}[x,y,0] = \left(\frac{2Re(x\bar{y})}{|x|^2 + |y|^2}, \frac{2Im(x\bar{y})}{|x|^2 + |y|^2}, \frac{|x|^2 - |y|^2}{|x|^2 + |y|^2}, \right).$$

ϕ is continuous since it is the composition of the continuous map

$$(u,v,w) \mapsto (u+iv, 1-w, 0)$$

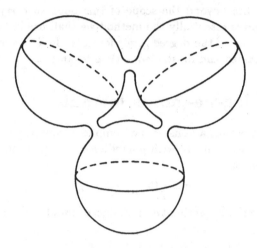

Figure 4.3: A deformation of a singular curve

from S^2 to $\mathbf{C}^3 - \{0\}$ with the map $\Pi : \mathbf{C}^3 - \{0\} \to \mathbf{P_2}$ defined by

$$\Pi(x, y, z) = [x, y, z],$$

while ϕ^{-1} is continuous since its composition with the restriction of Π to $\Pi^{-1}(L) \subseteq \mathbf{C}^3 - \{0\}$ is continuous (cf. remark 2.16). $\square$

This lemma shows that topologically our singular curve C_0 is homeomorphic to a union of d spheres meeting in $\frac{1}{2}d(d-1)$ points.

Now it seems intuitively reasonable that it is in fact always possible to perturb the coefficients of the polynomial defining C_0 by an arbitrarily small amount to get a *nonsingular* projective curve C_1. An equivalent statement is that if $\mathbf{C}_d[x, y, z]$ denotes the space of homogeneous polynomials of degree d in x, y, z with complex coefficients then the subset $\mathbf{C}_d^{\text{nonsing}}[x, y, z]$ of $\mathbf{C}_d[x, y, z]$ consisting of polynomials which define nonsingular curves is dense. This follows from the fact that the subset of $\mathbf{C}_d[x, y, z]$ consisting of polynomials which define singular curves has complex dimension one less than that of $\mathbf{C}_d[x, y, z]$. Note that $\mathbf{C}_d[x, y, z]$ can be identified with $\mathbf{C}^{\frac{1}{2}(d+1)(d+2)}$ by simply considering coefficients of $x^i y^j z^k$ where $i + j + k = d$.

It turns out that what happens on a topological level when such a perturbation takes place is that the singular points of intersection of the projective lines making up C_0 become very thin smooth "necks" or handles joining the projective lines (cf. figure 4.3).Precisely $d-1$ of these handles are used to join the d spheres together to form one sphere (topologically). Thus the non-singular perturbation C_1 of C_0 is topologically equivalent to a sphere with

$$\frac{1}{2}d(d-1) - (d-1) = \frac{1}{2}(d-1)(d-2)$$

handles. Thus we have at least one nonsingular curve with the right topology.

The second step in this method of proof is to show that if the coefficients of the polynomial defining a *nonsingular* curve are perturbed by a sufficiently small amount then the topology of the curve remains unchanged. This is perhaps plausible on an intuitive level but we shall not attempt to prove it.

The third step is to show that the space $C_d^{\mathrm{nonsing}}[x,y,z]$ of homogeneous polynomials of degree d in x,y,z defining nonsingular projective curves is path-connected. That is, given any two nonsingular projective curves of degree d in $\mathbf{P}_2$ defined by homogeneous polynomials $P(x,y,z)$ and $Q(x,y,z)$, there are nonsingular curves C_t defined by polynomials $P_t(x,y,z)$ for each $t \in [0,1]$ depending continuously on t such that $P_0(x,y,z) = P(x,y,z)$ and $P_1(x,y,z) = Q(x,y,z)$. The continuous map from $[0,1]$ to $C_d^{\mathrm{nonsing}}[x,y,z]$ defined by

$$t \mapsto P_t$$

is called a path in $C_d^{\mathrm{nonsing}}[x,y,z]$ from $P(x,y,z)$ to $Q(x,y,z)$. It is not difficult to show that such paths exist. If it were not for the requirement that each $P_t(x,y,z)$ should define a nonsingular curve we could just take

$$P_t(x,y,z) = (1-t)P(x,y,z) + tQ(x,y,z)$$

to get a path from $P(x,y,z)$ to $Q(x,y,z)$ in the space $C_d[x,y,z]$. Since such a path has real dimension one and the complement of $C_d^{\mathrm{nonsing}}[x,y,z]$ in $C_d[x,y,z]$ has real codimension two we can always shift the path very slightly to ensure that it misses the complement of $C_d^{\mathrm{nonsing}}[x,y,z]$ and hence defines a path in $C_d^{\mathrm{nonsing}}[x,y,z]$ as required.

Now we can put these three steps together to obtain the result we want. Let C be any nonsingular projective curve of degree d in $\mathbf{P}_2$. By the first step there exists a nonsingular projective curve C_1 of degree d in $\mathbf{P}_2$ which is topologically a sphere with

$$g = \frac{1}{2}(d-1)(d-2)$$

handles. By the third step there is a path

$$t \mapsto C_t, \quad t \in [0,1]$$

from C to C_1 in the space of nonsingular projective curves of degree d in $\mathbf{P}_2$. By the second step, given $t \in [0,1]$ there exists $\epsilon(t) > 0$ such that if $s \in [0,1]$ and $|t - s| < \epsilon(t)$ then C_s is homeomorphic to C_t. It is now easy to deduce that C_s and C_t are topologically equivalent for all $s, t \in [0,1]$. In particular C is homeomorphic to C_1 and hence is topologically a sphere with

$$g = \frac{1}{2}(d-1)(d-2)$$

handles.

4.1.2   The second method of proof

This method is related to the study of multivalued holomorphic functions. These have already been discussed briefly in 1.2.3 and they will make another appearance later: see §7.1.

Let
$$C = \{[x,y,z] \in \mathbf{P}_2 \; : \; P(x,y,z) = 0\}$$

be a nonsingular curve in $\mathbf{P}_2$. We may assume that C does not contain the point $[0,1,0]$, so that the coefficient of y^d in $P(x,y,z)$ is nonzero. We then put $z = 1$ to find the corresponding affine curve defined by the equation

$$P(x,y,1) = 0.$$

We shall regard this equation as defining y as a multivalued function of x.

Example 4.2 Let us consider the nonsingular curve C defined by

$$x^3 + y^3 + z^3 = 3yz^2.$$

Putting $z = 1$ we obtain the equation

$$x^3 + y^3 + 1 = 3y$$

which we can regard as defining y as a multivalued function of x; in fact

$$y = \left(\frac{-(x^3+1) + \sqrt{x^6 + 2x^3 - 3}}{2}\right)^{\frac{1}{3}} + \left(\frac{-(x^3+1) - \sqrt{x^6 + 2x^3 - 3}}{2}\right)^{\frac{1}{3}}$$

for appropriate choices of the cube roots (see [Stewart 73] pp. 161-163). Note that

$$x^6 + 2x^3 - 3 = 0$$

if and only if x^3 equals 1 or -3; i.e. x equals $1, \omega, \bar{\omega}, -\sqrt[3]{3}, -\omega\sqrt[3]{3}$ or $-\bar{\omega}\sqrt[3]{3}$ where $\omega = e^{\frac{2\pi i}{3}}$. To any value of x in

$$\mathbf{C} - \{1, \omega, \bar{\omega}, -\sqrt[3]{3}, -\omega\sqrt[3]{3}, -\bar{\omega}\sqrt[3]{3}\}$$

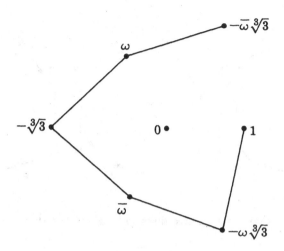

Figure 4.4: The cut plane D

there correspond exactly three values of y, and locally these define three holomorphic functions of x called "branches" of the multivalued function $y(x)$. When x travels around any of the points

$$\{1, \omega, \bar{\omega}, -\sqrt[3]{3}, -\omega\sqrt[3]{3}, -\bar{\omega}\sqrt[3]{3}\}$$

the value of y changes from one branch of the multivalued function to another. If we cut the complex plane $\mathbf{C}$ along the straight line segment $[1, -\omega\sqrt[3]{3}]$ from 1 to $-\omega\sqrt[3]{3}$ and along the straight line segments $[-\omega\sqrt[3]{3}, \bar{\omega}], [\bar{\omega}, -\sqrt[3]{3}], [-\sqrt[3]{3}, \omega]$ and $[\omega, -\bar{\omega}\sqrt[3]{3}]$ (see figure 4.4)then we find that there are three (single-valued) holomorphic functions f_1, f_2, f_3 defined on the cut plane D satisfying

$$f_j(x)^3 + x^3 + 1 = 3f_j(x)$$

for $j = 1, 2, 3$, and all $x \in D$. Therefore the subset

$$\{[x, y, 1] \in C \;:\; x \in D\} = \bigcup_{j=1}^{3}\{[x, y, 1] \in \mathbf{P}_2 \;:\; x \in D, y = f_j(x)\}$$

of C is the disjoint union of three copies of the cut plane D. If we add in the three points at infinity this means that we can construct C topologically by taking three copies of $D \cup \{\infty\}$ and glueing the edges of the cuts together appropriately, corresponding to the way the multivalued function jumps from one branch f_j to another as y crosses the cuts.

Note that each copy of $D \cup \{\infty\}$ is topologically a disc, and its boundary is made up of ten line segments. One can check that the boundaries should

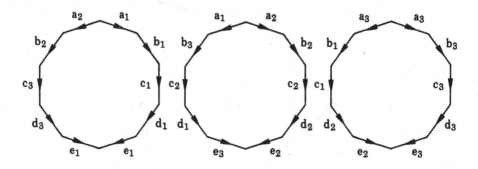

Figure 4.5: Three discs to be glued together

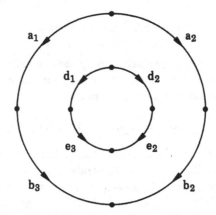

Figure 4.6: Annulus formed from second disc

be identified according to figure 4.5, in which a_j, b_j, c_j, d_j, e_j refer to cuts
along $[1, -\omega\sqrt[3]{3}], [-\omega\sqrt[3]{3}, \bar{\omega}], [\bar{\omega}, -\sqrt[3]{3}], [-\sqrt[3]{3}, \omega]$ and $[\omega, -\bar{\omega}\sqrt[3]{3}]$ respectively.
The two copies of c_2 in the second disc can be identified to give an annulus
whose boundary is given by figure 4.6. The two pairs b_1c_1 and c_3d_3 occuring in
the first and third discs in figure 4.5 can be identified to give another annulus
whose boundary is given by figure 4.7. Note that the adjacent copies of a_3
and e_1 on this figure can be glued together and thus eliminated. Hence the
two annuli given in figures 4.6 and 4.7 can be glued together to give a sphere
with one handle, or torus (figure 4.8).
This agrees with the degree-genus formula

$$g = \frac{1}{2}(d-1)(d-2);$$

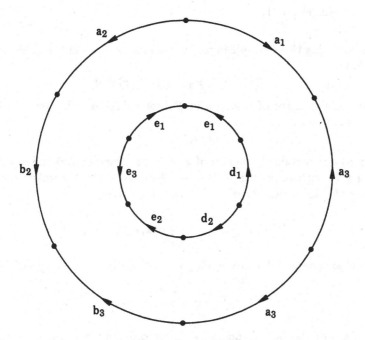

Figure 4.7: Annulus formed from first and third discs

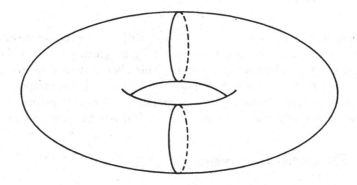

Figure 4.8: Torus

when $d = 3$ we get $g = 1$.

After studying this example in some detail let us consider the general case again. So let

$$C = \{[x, y, z] \in \mathbf{P_2} : P(x, y, z) = 0\}$$

be a nonsingular curve of degree d in $\mathbf{P_2}$ not containing the point $[0, 1, 0]$. Then the equation

$$P(x, y, 1) = 0$$

defines y as a multivalued function of x such that there correspond exactly d values of y to each value of $x \in C$ other than the "branch points", that is, the values of x for which there is a value of y satisfying

$$P(x, y, 1) = 0 = \frac{\partial P}{\partial y}(x, y, 1).$$

The point ∞ is regarded as a branch point if there is a value of y satisfying

$$P(1, y, 0) = 0 = \frac{\partial P}{\partial y}(1, y, 0).$$

If $p_1, \ldots, p_r$ are the branch points, suitably ordered, then we can cut the complex plane $\mathbf{C}$ along the straight line segments

$$[p_1, p_2], [p_2, p_3], \ldots, [p_{r-1}, p_r]$$

and define holomorphic functions $f_1, \ldots, f_d$ on the cut plane such that if y lies in the cut plane then

$$P(x, y, 1) = 0$$

if and only if $y = f_j(x)$ for some $j \in \{1, \ldots, d\}$. Then just as in example 4.2 we can construct the curve C topologically by glueing together d copies of the cut complex plane along the cuts. This always gives a sphere with a certain number of handles (see Appendix C). However the computation of this number g requires some more information about branch points. For this reason we shall study branch points more carefully in the next section.

4.2   Branched covers of $\mathbf{P}_1$

The nonsingular projective curve C in $\mathbf{P_2}$ defined by the equation $y^2 = xz$ admits a surjection $\phi : C \to \mathbf{P}_1$ defined by $\phi[x, y, z] = [x, z]$ such that if $[x, z] \in \mathbf{P}_1$ then $\phi^{-1}([x, z])$ consists of exactly two points unless $x = 0$ or $z = 0$. Such a map $\phi : C \to \mathbf{P}_1$ is called a *double cover* of $\mathbf{P}_1$ *branched* over the points $[0, 1]$ and $[1, 0]$. The points in C mapping to $[0, 1]$ and $[1, 0]$ are called *ramification points* of ϕ. We can use ϕ to visualise C as two copies of $\mathbf{P_1}$ cut and glued together as in §1.2.3 (cf. also example 4.2).

We shall see in this section that any nonsingular projective curve C of degree $d > 1$ in $\mathbf{P_2}$ can be viewed as a branched cover of $\mathbf{P_1}$ in a similar way. We shall show that if we choose coordinates on $\mathbf{P_2}$ appropriately then the number of ramification points is precisely $d(d-1)$. We shall use this in the next section to prove the degree-genus formula.

Let C be a nonsingular projective curve in $\mathbf{P_2}$ defined by a homogeneous polynomial $P(x, y, z)$ of degree $d > 1$. By applying a suitable projective transformation we may assume that $[0, 1, 0] \notin C$. Then we have a well-defined map $\phi : C \to \mathbf{P_1}$ given by

$$\phi[x, y, z] = [x, z].$$

Definition 4.3 *The* ramification index $\nu_\phi[a, b, c]$ *of* ϕ *at a point* $[a, b, c] \in C$ *is the order of the zero of the polynomial* $P(a, y, c)$ *in* y *at* $y = b$. *The point* $[a, b, c]$ *is called a* ramification point *of* ϕ *if* $\nu_\phi[a, b, c] > 1$.

Remark 4.4
(i) $\nu_\phi[a, b, c] > 0$ if and only if $[a, b, c] \in C$.
(ii) $\nu_\phi[a, b, c] > 1$ if and only if

$$P(a, b, c) = 0 = \frac{\partial P}{\partial y}(a, b, c),$$

i.e. if and only if $[a, b, c] \in C$ and the tangent line to C at $[a, b, c]$ contains the point $[0, 1, 0]$.
(iii) $\nu_\phi[a, b, c] > 2$ if and only if

$$P(a, b, c) = 0 = \frac{\partial P}{\partial y}(a, b, c) = \frac{\partial^2 P}{\partial y^2}(a, b, c).$$

This happens if and only if $[a, b, c]$ is a *point of inflection* on C and the tangent line to C at $[a, b, c]$ contains the point $[0, 1, 0]$. To prove this, note that $[a, b, c] \neq [0, 1, 0]$ so $a \neq 0$ or $c \neq 0$. Let us assume that $c \neq 0$; the case $a \neq 0$ is similar. Suppose

$$P(a, b, c) = 0 = \frac{\partial P}{\partial y}(a, b, c);$$

then by lemma 3.30 we have

$$\mathcal{H}_P(a, b, c) = \frac{(d-1)^2}{c^2} \det \begin{pmatrix} P_{xx} & P_{xy} & P_x \\ P_{yx} & P_{yy} & 0 \\ P_x & 0 & 0 \end{pmatrix}$$

$$= -\frac{(d-1)^2}{c^2}(P_x)^2 P_{yy}$$

where the partial derivatives are evaluated at (a, b, c). By Euler's relation (lemma 2.32) if

$$\frac{\partial P}{\partial x}(a, b, c) = 0$$

then as $c \neq 0$ and $\frac{\partial P}{\partial y}(a, b, c) = 0$ the point $[a, b, c] \in C$ would be singular, which contradicts the assumption that C is a nonsingular curve. Hence $\mathcal{H}_P(a, b, c) = 0$ if and only if

$$\frac{\partial^2 P}{\partial y^2}(a, b, c) = 0$$

as required.

Lemma 4.5 *The inverse image $\phi^{-1}([a, c])$ of any $[a, c]$ in $\mathbf{P}_1$ under ϕ contains exactly*

$$d - \sum_{p \in \phi^{-1}([a,c])} (\nu_\phi(p) - 1)$$

points. In particular $\phi^{-1}([a, c])$ contains d points if and only if $\phi^{-1}([a, c])$ contains no ramification points of ϕ.

Proof. A point of C lies in $\phi^{-1}([a, c])$ if and only if it is of the form $[a, b, c]$ where $b \in \mathbf{C}$ satisfies

$$P(a, b, c) = 0.$$

By assumption $[0, 1, 0] \notin C$ so $P(0, 1, 0) \neq 0$. Hence we may assume that $P(0, 1, 0) = 1$. Then $P(a, y, c)$ is a monic polynomial of degree d in y so

$$P(a, y, c) = \prod_{1 \leq i \leq r} (y - b_i)^{m_i}$$

where $b_1, \ldots, b_r$ are distinct complex numbers and $m_1, \ldots, m_r$ are positive integers such that

$$m_1 + \ldots + m_r = d.$$

Thus

$$\phi^{-1}([a, c]) = \{[a, b_i, c] \ : \ 1 \leq i \leq r\}$$

and the ramification index of ϕ at $[a, b_i, c]$ is

$$\nu_\phi[a, b_i, c] = m_i.$$

The result follows. $\square$

Definition 4.6 *Let R be the set of ramification points of ϕ. The image $\phi(R)$ of R under ϕ is called the* branch locus *of ϕ, and $\phi : C \to \mathbf{P}_1$ is called a* branched cover *of $\mathbf{P}_1$.*

Lemma 4.7 *(i)* ϕ *has at most* $d(d-1)$ *ramification points.*
(ii) If $\nu_\phi[a, b, c] \le 2$ *for all* $[a, b, c] \in C$ *then* C *has exactly* $d(d-1)$ *ramification points.*

Proof. Since C is nonsingular it is irreducible (see corollary 3.10). By assumption $[0, 1, 0] \notin C$ so the coefficient $P(0, 1, 0)$ of y^d in $P(x, y, z)$ is nonzero. Thus the homogeneous polynomial

$$\frac{\partial P}{\partial y}(x, y, z)$$

is not identically zero and has degree $d - 1$, so it cannot be divisible by $P(x, y, z)$. Hence the projective curve D of degree $d - 1$ defined by this polynomial has no component in common with C. Thus (i) follows from the weak form (theorem 3.9) of Bézout's theorem, because the set R of ramification points of C is the intersection of C and D.

Now suppose that

$$\nu_\phi[a, b, c] \le 2$$

for all $[a, b, c] \in C$. By the corollary 3.25 to the strong form of Bézout's theorem, in order to prove (ii) it suffices to show that if $[a, b, c]$ lies in $R = C \cap D$ then $[a, b, c]$ is a nonsingular point of D and the tangent lines to C and D at $[a, b, c]$ are distinct. If not then $[a, b, c]$ satisfies

$$P(a, b, c) = 0 = P_y(a, b, c)$$

because it lies in C and D, and the vector

$$(P_{xy}(a, b, c), P_{yy}(a, b, c), P_{zy}(a, b, c))$$

is either zero or a scalar multiple of the vector

$$(P_x(a, b, c), P_y(a, b, c), P_z(a, b, c)).$$

This implies that

$$P(a, b, c) = 0 = P_y(a, b, c) = P_{yy}(a, b, c),$$

i.e.

$$\nu_\phi[a, b, c] > 2.$$

This contradiction completes the proof. $\square$

Lemma 4.8 *By applying a suitable projective transformation to* C *we may assume that*

$$\nu_\phi[a, b, c] \le 2$$

for all $[a, b, c] \in C.$

Proof. By proposition 3.33 C has only a finite number (at most $3d(d-2)$) of points of inflection. Thus by applying a suitable projective transformation we may assume that $[0, 1, 0]$ does not lie on C nor on any of the tangent lines to C at its points of inflection. The result now follows from remark 4.4(iii). $\square$

4.3   Proof of the degree-genus formula

We can now use the results on branched covers in the last section to prove the degree-genus formula

$$g = \frac{1}{2}(d-1)(d-2)$$

relating the degree d and the genus g of a nonsingular complex projective curve in $\mathbf{P}_2$. The first task is to give a precise definition of the term genus. For this purpose we introduce the concept of a triangulation of a curve C — the rough idea is that we divide C up into triangles. Let

$$\Delta = \{(x, y) \in \mathbf{R}^2 \; : \; x \geq 0, y \geq 0, x + y \leq 1\}$$

be the standard triangle in $\mathbf{R}^2$ with vertices $(0, 0), (1, 0)$ and $(0, 1)$. Let

$$\Delta^0 = \{(x, y) \in \mathbf{R}^2 \; : \; x > 0, y > 0, x + y < 1\}$$

be its interior.

Definition 4.9 *Let C be a nonsingular complex projective curve in $\mathbf{P}_2$. A triangulation of C is given by the following data:*
(a) a finite nonempty set V of points called vertices,
(b) a finite nonempty set E of continuous maps $e : [0, 1] \to C$ called edges,
(c) a finite nonempty set F of continuous maps $f : \Delta \to C$ called faces,
satisfying
(i) $V = \{e(0) : e \in E\} \cup \{e(1) : e \in E\}$, i.e. the vertices are the endpoints of the edges;
(ii) if $e \in E$ then the restriction of e to the open interval $(0, 1)$ is a homeomorphism onto its image in C, and this image contains no points in V or in the image of any other edge $\tilde{e} \in E$;
(iii) if $f \in F$ then the restriction of f to Δ^0 is a homeomorphism onto a connected component K_f of $C - \Gamma$ where

$$\Gamma = \bigcup_{e \in E} e([0, 1])$$

is the union of the images of the edges, and if $r : [0,1] \to [0,1]$ *and* $\sigma_i :$ $[0,1] \to \Delta$ *for* $1 \le i \le 3$ *are defined by*

$$r(t) = 1 - t, \quad \sigma_1(t) = (t,0), \quad \sigma_2(t) = (1-t,t), \quad \sigma_3(t) = (0,1-t)$$

then either $f \circ \sigma_i$ *or* $f \circ \sigma_i \circ r$ *is an edge* $e_f^i \in E$ *for* $1 \le i \le 3$;
(iv) the mapping $f \mapsto K_f$ *from* F *to the set of connected components of* $C - \Gamma$
is a bijection;
(v) for every $e \in E$ *there is exactly one face* $f_e^+ \in F$ *such that* $e = f_e^+ \circ \sigma_i$ *for*
some $i \in \{1,2,3\}$ *and exactly one face* $f_e^- \in F$ *such that* $e = f_e^- \circ \sigma_i \circ r$ *for*
some $i \in \{1,2,3\}$.

Remark 4.10 This is not quite the standard definition of a triangulation but it is convenient for our purposes. In particular we can only assume that V, E and F are finite since we are dealing with a *compact* space. In addition condition 4.9(v) tells us that the triangulation is "coherently oriented" (cf. [Springer 57] §5.3).

Remark 4.11 It is common to confuse edges and faces with their images in C.

Definition 4.12 *The* Euler number χ *of a triangulation is defined by*

$$\chi = \#V - \#E + \#F$$

where the symbol $\#S$ *denotes the number of elements of a finite set* S.

The Euler number is important because of the following theorem.

Theorem 4.13 *(i) Every nonsingular projective curve* C *in* $\mathbf{P}_2$ *has a triangulation.*
(ii) The Euler number χ *of a triangulation of* C *depends only on* C, *not on the triangulation.*

Because of this theorem we can define the Euler number $\chi(C)$ of C to be the Euler number of any triangulation of C.

Examples 4.14 (i) By lemma 4.1 a complex projective line in $\mathbf{P}_2$ is homeomorphic to a sphere. Thus it has a triangulation with three vertices, three edges and two faces (see figure 4.9). Therefore

$$\chi(\mathbf{P}_1) = 3 - 3 + 2 = 2.$$

(ii) The nonsingular cubic curve C defined by

$$y^2 z = x(x - z)(x - \lambda z)$$

with $\lambda \ne 0,1$ is topologically a torus (see §1.2.3) so it has a triangulation with one vertex, three edges, and two faces (see figure 4.10). Thus

$$\chi(C) = 1 - 3 + 2 = 0.$$

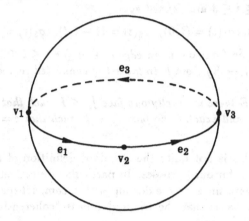

Figure 4.9: A triangulation of a sphere

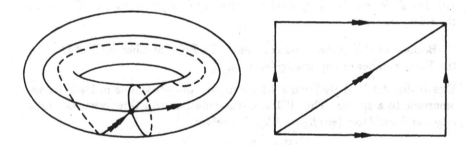

Figure 4.10: A triangulation of the torus

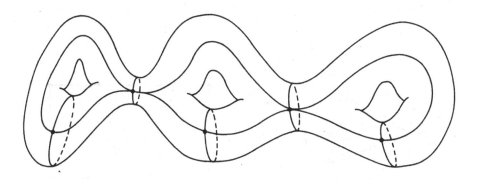

Figure 4.11: A sphere with three handles

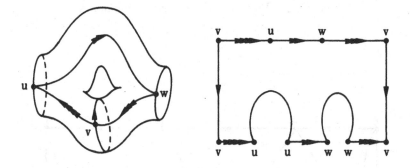

Figure 4.12: Piece of a sphere with handles

Definition 4.15 *The genus of a nonsingular projective curve C is*

$$g = \frac{1}{2}(2 - \chi)$$

where χ is the Euler number of C.

Example 4.16 Suppose that C is homeomorphic to a sphere with g handles (see figure 4.11 for a picture when $g = 3$). Then C can be cut into $g - 2$ pieces of the form illustrated in figure 4.12 together with two endpieces of the form illustrated in figure 4.13. These pieces can be subdivided into triangles, for example as in figures 4.14 and 4.15.

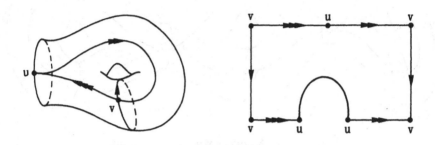

Figure 4.13: Endpiece of a sphere with handles

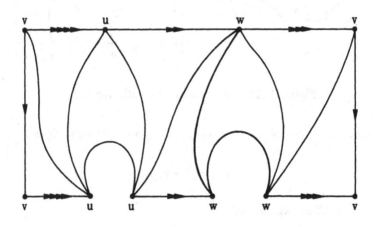

Figure 4.14: A subdivision of figure 4.12

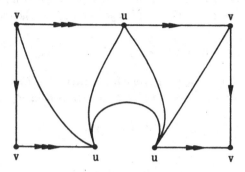

Figure 4.15: A subdivision of figure 4.13

Thus C has a triangulation with $2g - 1$ vertices, $12g - 9$ edges and $8g - 6$ faces, so its Euler number is

$$\chi = (2g - 1) - (12g - 9) + 8g - 6 = 2 - 2g$$

and the genus of C is

$$\frac{1}{2}(2 - \chi) = g.$$

This shows that definition 4.15 of the genus of C fits with our original description of the genus as the "number of handles".

Remark 4.17 If $h : C \to D$ is a homeomorphism between two nonsingular projective curves C and D, and if C has a triangulation with V, E, F as the sets of vertices, edges and faces, then D has a triangulation with vertices

$$\{h(v) \; : \; v \in V\},$$

edges

$$\{h \circ e \; : \; e \in E\},$$

and faces

$$\{h \circ f \; : \; f \in F\}.$$

Thus C and D have the same Euler number and genus. In other words the Euler number and genus of a nonsingular projective curve are *topological invariants*: they only depend on the topology of the curve, not its algebraic structure.

A proof of theorem 4.13(ii) is given in Appendix C. As for theorem 4.13(i), we have the following more detailed result.

Theorem 4.18 *Let C be a nonsingular projective curve of degree d in $\mathbf{P}_2$. If r is a positive integer and $r \geq d(d-1)$ and $r \geq 3$ then C has a triangulation with $rd - d(d-1)$ vertices, $3(r-2)d$ edges and $2(r-2)d$ faces.*

This theorem has the following immediate corollary.

Corollary 4.19 (The degree-genus formula) *The Euler number χ and genus g of a nonsingular projective curve of degree d in $\mathbf{P}_2$ are given by*

$$\chi = d(3 - d)$$

and

$$g = \frac{1}{2}(d-1)(d-2).$$

Examples 4.20 (1) When $d = 1$ then $\chi = 2$ and $g = 0$. A projective curve of degree 1 in $\mathbf{P}_2$ is a line so this tallies with example 4.14(i).
(2) When $d = 2$ then $\chi = 2$ and $g = 0$ again. A projective curve of degree 2 in $\mathbf{P}_2$ is a conic, and by remark 3.13 any nonsingular projective conic is homeomorphic to $\mathbf{P}_1$, or equivalently to a complex projective line in $\mathbf{P}_2$, so this tallies with example 4.14(i).
(3) When $d = 3$ then $\chi = 0$ and $g = 1$. By corollary 3.34 a nonsingular projective cubic curve in $\mathbf{P}_2$ is equivalent under a projective transformation to the curve defined by

$$y^2 z = x(x - z)(x - \lambda z)$$

for some $\lambda \in \mathbf{C} - \{0, 1\}$, so this tallies with example 4.14(ii).

For the proof of theorem 4.18 we need two preliminary results.

Lemma 4.21 *Let $\{p_1, \ldots, p_r\}$ be any set of least three points in $\mathbf{P}_1$. Then there is a triangulation of $\mathbf{P}_1$ with $p_1, \ldots, p_r$ as its vertices and with $3r - 6$ edges and $2r - 4$ faces.*

Proposition 4.22 *Let*

$$C = \{[x, y, z] \in \mathbf{P}_2 \ : \ P(x, y, z) = 0\}$$

be a nonsingular projective curve not containing $[0, 1, 0]$ and let $\phi : C \to \mathbf{P}_1$ be the branched cover defined by $\phi[x, y, z] = [x, z]$. Suppose that (V, E, F) is a triangulation of $\mathbf{P}_1$ such that the set of vertices V contains the branch locus $\phi(R)$ of ϕ. Then there is a triangulation $(\tilde{V}, \tilde{E}, \tilde{F})$ of C such that

$$\tilde{V} = \phi^{-1}(V),$$

$$\tilde{E} = \{\tilde{e} : [0,1] \to C \ : \ \tilde{e} \text{ continuous}, \phi \circ \tilde{e} \in E\}$$

and

$$\tilde{F} = \{\tilde{f} : \Delta \to C \ : \ \tilde{f} \text{ continuous}, \phi \circ \tilde{f} \in F\}.$$

Moreover if $\nu_\phi(p)$ is the ramification index of ϕ at p and d is the degree of C then

$$\#\tilde{V} = d\#V - \sum_{p \in R}(\nu_\phi(p) - 1),$$

$$\#\tilde{E} = d\#E$$

and

$$\#\tilde{F} = d\#F.$$

Remark 4.23 It follows from this proposition that the Euler number $\chi(C)$ of C is given by

$$\chi(C) = \#\tilde{V} - \#\tilde{E} + \#\tilde{F} = d\chi(\mathbf{P}_1) - \sum_{p \in R}(\nu_\phi(p) - 1).$$

This is called the *Riemann-Hurwitz formula* for the branched cover $\phi : C \to \mathbf{P}_1$.

Proof of theorem 4.18 given 4.21 and 4.22. Let $P(x,y,z)$ be a homogeneous polynomial of degree d defining the curve C. By lemma 4.8, after applying a suitable projective transformation to C we may assume that the map $\phi : C \to \mathbf{P}_1$ given by

$$\phi[x,y,z] = [x,z]$$

is well-defined (i.e. $[0,1,0] \notin C$) and the ramification index $\nu_\phi[a,b,c]$ of ϕ at every $[a,b,c] \in C$ satisfies

$$\nu_\phi[a,b,c] \le 2.$$

Then by lemma 4.7 ϕ has exactly $d(d-1)$ ramification points, i.e. $\#R = d(d-1)$.

By lemma 4.21 if $r \ge 3$ and $r \ge d(d-1)$ then we can choose a triangulation (V, E, F) of $\mathbf{P}_1$ such that

$$V \supseteq \phi(R)$$

and $\#V = r, \#E = 3r - 6$ and $\#F = 2r - 4$. Therefore by proposition 4.22 there is a triangulation $(\tilde{V}, \tilde{E}, \tilde{F})$ of C with

$$\#\tilde{E} = d\#E = 3(r-2)d,$$

$$\#\tilde{F} = d\#F = 2(r-2)d$$

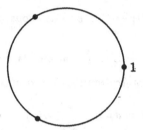

Figure 4.16: A subdivision of $\mathbf{C}$

and

$$\#\tilde{V} = d\#V - \sum_{p \in R}(\nu_\phi(p) - 1).$$

Since $\#R = d(d-1)$ and $\nu_\phi(p) = 2$ for all $p \in R$ we have

$$\#\tilde{V} = rd - d(d-1)$$

as required. $\square$

Proof of lemma 4.21. We can prove this for example by induction on $r \geq 3$. When $r = 3$ lemma 2.22 tells us that there is a projective transformation taking p_1 to 1, p_2 to $e^{\frac{2\pi i}{3}}$ and p_3 to $e^{\frac{4\pi i}{3}}$. (Here as usual we identify $\mathbf{P}_1$ with $\mathbf{C} \cup \{\infty\}$.) We can join these three points by segments of the unit circle in $\mathbf{C}$ (see figure 4.16).

The exterior of the unit circle together with the point at infinity is mapped by the projective transformation $z \mapsto \frac{1}{z}$ to the interior of the unit circle. Since there is a homeomorphism

$$\triangle \to \{z \in \mathbf{C} : |z| \leq 1\}$$

which takes the vertices of the triangle $\triangle$ to $1, e^{\frac{2\pi i}{3}}$ and $e^{\frac{4\pi i}{3}}$ and takes the edges of $\triangle$ to the appropriate segments of the unit circle, we get a triangulation of $\mathbf{P}_1$ with three edges and two faces when $r = 3$.

Now suppose $r > 3$ and we have a triangulation with vertices $p_1, \ldots, p_{r-1}$ and $3r - 9$ edges and $2r - 6$ faces. If p_r lies in the interior of a face f (more precisely, if $p_r \in f(\triangle^0)$) we can add three edges joining p_r to the vertices of the face f (see figure 4.17) and obtain a new triangulation with one extra vertex, p_r, three extra edges and one old face subdivided into three new ones. If p_r does not lie in the interior of a face then it lies on an edge e (more precisely $p_r = e(t)$ for some $t \in (0,1)$). We can then replace e by two edges joining p_r to $e(0)$ and $e(1)$, and add two edges joining p_r to the remaining vertices of the faces f_e^+ and f_e^- (see figure 4.18). This gives a triangulation with one new vertex, p_r, two extra edges and one old edge replaced by two new

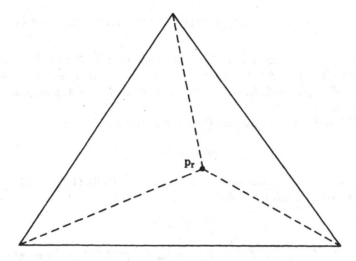

Figure 4.17: A subdivision of a triangle

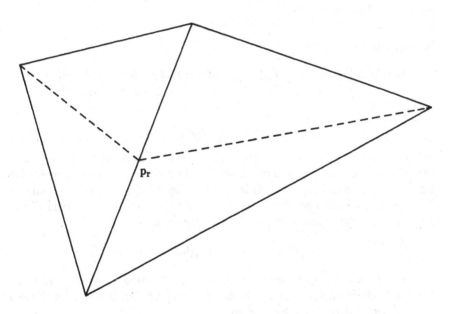

Figure 4.18: A subdivision of two adjacent triangles

ones, and two old faces each replaced by two new ones. The result follows. □

Proof of proposition 4.22. We must show that $\tilde{V}, \tilde{E}$ and $\tilde{F}$ satisfy the conditions (i)–(v) in the definition 4.9 of a triangulation, and that the formulas for $\#\tilde{V}, \#\tilde{E}$ and $\#\tilde{F}$ are correct. This uses some topological results proved in Appendix C.

By C.8 and C.9 of Appendix C, if $f \in F$ and $p \in C$ and

$$\phi(p) = f(t)$$

for some $t \in \Delta$ not equal to any of the vertices $(0,0), (1,0)$ or $(0,1)$, then there is a unique continuous map

$$\tilde{f} : \Delta \to C$$

such that $\phi \circ \tilde{f} = f$ and $\tilde{f} = p$. By lemma 4.5 $\phi^{-1}\{f(t)\}$ consists of exactly d points of C (because $f(t)$ does not belong to the branch locus $\phi(R)$) so we can deduce that there are exactly d continuous maps $\tilde{f} : \Delta \to C$ such that $\phi \circ \tilde{f} = f$. This means that

$$\#\tilde{F} = d\#F.$$

We can also deduce that

$$
\begin{aligned}
C - \phi^{-1}(V) &= \phi^{-1}\{f(t) \ : \ f \in F, t \in \Delta, t \neq (0,0), (1,0), (0,1)\} \\
&= \{\tilde{f}(T) \ : \ \tilde{f} \in \tilde{F}, t \in \Delta, t \neq (0,0), (1,0), (0,1)\}.
\end{aligned}
$$

In particular

$$G = \cup_{\tilde{f} \in \tilde{F}}\,\tilde{f}(\Delta)$$

contains $C - \phi^{-1}(V)$ and is therefore dense in C because $\phi^{-1}(V)$ is finite by lemma 4.5. (The complement of a finite set of points in C is always dense: see e.g. remark 5.29 in §5.2.) But Δ is compact (by 2.12(i)) so $\tilde{f}(\Delta)$ is compact if $\tilde{f} \in \tilde{F}$ (by 2.12(ii)) so G is compact (by 2.12(vi)) so G is closed in C (by 2.12(v) and 2.30). Thus $G = C$, which implies that

$$\phi^{-1}(V) = \{\tilde{f}(t) \ : \ \tilde{f} \in \tilde{F}, t \in \{(0,0), (1,0), (0,1)\}\}.$$

If $\tilde{f} \in \tilde{F}$ then $\phi \circ \tilde{f} \in F$ so either $\phi \circ \tilde{f} \circ \sigma_i \in E$ or $\phi \circ \tilde{f} \circ \sigma_i \circ r \in E$ for $1 \le i \le 3$ where $\sigma_1, \sigma_2, \sigma_3$ and r are defined as in 4.9(iii). Thus if $\tilde{f} \in \tilde{F}$ then either $\tilde{f} \circ \sigma_i \in \tilde{E}$ or $\tilde{f} \circ \sigma_i \circ r \in \tilde{E}$ so

$$\tilde{f}(t) \in \{\tilde{e}(0) \ : \ \tilde{e} \in \tilde{E}\} \cup \{\tilde{e}(1) \ : \ \tilde{e} \in \tilde{E}\}$$

if $t \in \{(0,0), (1,0), (0,1)\}$. This tells us that

$$\phi^{-1}(V) = \{\tilde{e}(0) \ : \ \tilde{e} \in \tilde{E}\} \cup \{\tilde{e}(1) \ : \ \tilde{e} \in \tilde{E}\};$$

i.e. that condition (i) of 4.9 is satisfied.

It follows from lemma C.7 and remark C.9 of Appendix C that if $e \in E$ and $p \in C$ and

$$\phi(p) = e(t)$$

for some $t \in (0,1)$ then there is a unique continuous map

$$\tilde{e} : [0,1] \to C$$

such that $\phi \circ \tilde{e} = e$ and $\tilde{e}(t) = p$; and moreover by lemma C.6 the restriction of $\tilde{e}$ to $(0,1)$ is a homeomorphism onto its image in C. Condition (ii) of definition 4.9 follows using the uniqueness of $\tilde{e}$. It also follows that

$$\phi^{-1}\{e(t) \; : \; e \in E, t \in (0,1)\} = \{\tilde{e}(t) \; : \; \tilde{e} \in \tilde{E}, t \in (0,1)\}.$$

Therefore if

$$\Gamma = \bigcup_{e \in E} e([0,1]) = V \cup \{e(t) \; : \; e \in E, t \in (0,1)\}$$

then

$$\phi^{-1}(\Gamma) = \phi^{-1}(V) \cup \{\tilde{e}(t) \; : \; \tilde{e} \in \tilde{E}, t \in (0,1)\} = \tilde{\Gamma}$$

where

$$\tilde{\Gamma} = \bigcup_{\tilde{e} \in \tilde{E}} \tilde{e}([0,1]).$$

Furthermore by lemma 4.5 if $t \in (0,1)$ and $e \in E$ then $\phi^{-1}\{e(t)\}$ consists of exactly d points of C (because $e(t)$ does not belong to the branch locus $\phi(R)$) so there are exactly d continuous maps $\tilde{e} : [0,1] \to C$ such that $\phi \circ \tilde{e} = e$. Thus

$$\#\tilde{E} = d\#E.$$

By lemma C.6 of Appendix C if $\tilde{f} \in \tilde{F}$ then the restriction of $\tilde{f}$ to Δ^0 is a homeomorphism onto its image which is a connected component of $\phi^{-1}(f(\Delta^0))$ where $f = \phi \circ \tilde{f}$. Since $f(\Delta^0)$ is a connected component of $\mathbf{P}_1 - \Gamma$ it follows that $\tilde{f}(\Delta^0)$ is a connected component of

$$\phi^{-1}(\mathbf{P}_1 - \Gamma) = C - \phi^{-1}(\Gamma) = C - \tilde{\Gamma}.$$

This shows that the first half of condition (iii) is satisfied; we have already noted that the second half is true. Conditions (iv) and (v) follow easily from what we have already done. Thus it remains to show that

$$\#\tilde{V} = d\#V - \sum_{p \in R}(\nu_\phi(p) - 1).$$

This follows immediately from lemma 4.5 since V contains $\phi(R)$. $\quad\square$

4.4   Exercises

4.1. Let C and D be nonsingular projective curves of degrees n and m in $\mathbf{P}_2$. Show that if C is homeomorphic to D then either $n = m$ or $\{n, m\} = \{1, 2\}$.

4.2. Let $\phi : C \to \mathbf{P}_1$ be defined by

$$\phi[x, y, z] = [x, z]$$

where C is a nonsingular projective curve in the projective plane not containing the point $[0, 1, 0]$. Show that if C has degree $d > 1$ then ϕ has at least one ramification point. Show that if $d = 1$ then ϕ has no ramification points and is a homeomorphism.

4.3. Show that the projective curve D defined by $y^2 z = x^3$ has a unique singular point. Show that the map $f : \mathbf{P}_1 \to D$ defined by

$$f[s, t] = [s^2 t, s^3, t^3]$$

is a homeomorphism. Deduce that the degree–genus formula cannot be applied to singular curves in $\mathbf{P}_2$.

4.4. Show that there is a homeomorphism given by

$$[s, t, 0] \mapsto [st^3, (s + t)^4, t^4]$$

from the line in $\mathbf{P}_2$ defined by $z = 0$ onto a quartic curve in $\mathbf{P}_2$. Why does this not contradict exercise 4.1?

4.5. Let C be an irreducible projective cubic curve in $\mathbf{P}_2$ with a singular point p. Using exercise 3.9 or otherwise, show that there is a continuous surjection $f : \mathbf{P}_1 \to C$ defined as follows. Identify $\mathbf{P}_1$ with the set of lines in $\mathbf{P}_2$ which pass through p; then map a line L through p to p if it is a tangent line to C at p, and otherwise to the unique other point of intersection of L with C.

Show that if C is a cuspidal cubic, that is, equivalent under a projective transformation to the curve defined by

$$y^2 z = x^3,$$

then f is a homeomorphism (cf. exercise 4.3), whereas if C is a nodal cubic, that is, equivalent to the curve defined by

$$y^2 z = x^2 (x + z)$$

then f maps two distinct points of $\mathbf{P}_1$ to p but is a homeomorphism away from p. [Hint: recall that the image of a closed subset of a compact space under a continuous map to a Hausdorff space is always closed, and in particular any continuous bijection from a compact space to a Hausdorff space is a homeomorphism].

Chapter 5

Riemann surfaces

The results of Chapter 4 imply that a nonsingular complex projective curve C in $\mathbf{P}_2$ is topologically a surface; that is, every point on C has a neighbourhood which is homeomorphic to an open subset of the Euclidean plane $\mathbf{R}^2$. However a nonsingular complex curve C is a very special sort of surface: it has much more structure than is given by its topology alone. In particular it makes sense to talk about holomorphic functions and do complex analysis on C, which means that C is what is called a *Riemann surface*.

We shall give a proper definition of a Riemann surface in §5.2. Before that we shall recall in §5.1 some of the important results and definitions we shall need from complex analysis, and we shall illustrate them using a function which will be important later: the Weierstrass $\wp$-function.

5.1 The Weierstrass $\wp$-function

Let W be an open subset of the complex plane $\mathbf{C}$. Recall that a function $f : W \to \mathbf{C}$ is called *holomorphic* if its derivative

$$f'(a) = \lim_{z \to a} \frac{f(z) - f(a)}{z - a}$$

exists at every $a \in W$. A function f is holomorphic on an open disc

$$\{z \in \mathbf{C} : |z - a| < r\}$$

centred at $a \in \mathbf{C}$ if and only if it can be expressed as a convergent power series

$$f(z) = \sum_{n \geq 0} c_n(z - a)^n, \quad |z - a| < r$$

111

called the Taylor series of f about a (see e.g. [Priestley 85] 5.9). The coefficient c_n is found by differentiating the power series n times at a: we have

$$c_n = \frac{1}{n!} f^{(n)}(a).$$

A *meromorphic* function on W is a function

$$f : W \to \mathbf{C} \cup \{\infty\}$$

such that $f : W - f^{-1}\{\infty\} \to \mathbf{C}$ is holomorphic and F has a *pole* at each $a \in f^{-1}\{\infty\}$: that is, near a we can write

$$f(z) = \frac{g(z)}{(z-a)^m}$$

for some $m > 0$ where $g(z)$ is holomorphic in an open neighbourhood of a and $g(a) \neq 0$. Equivalently we can express $f(z)$ as a Laurent series

$$f(z) = \sum_{n \geq -m} c_n (z-a)^n$$

where

$$g(z) = \sum_{n \geq 0} c_{n-m} (z-a)^n$$

is the Taylor expansion of $g(z)$ and

$$c_{-m} = g(a) \neq 0.$$

Then m is called the *order* or *multiplicity* of the pole. The coefficient c_{-1} is called the *residue* of $f(z)$ at a and is denoted by

$$res\{f(z); a\}$$

Remark 5.1 It is easy to check that in this situation

$$\frac{f'(z)}{f(z)}$$

has a simple pole (that is, a pole of order one) at a with residue $-m$. Similarly if f has a zero of multiplicity (or order) m at a, that is, if near a we can write

$$f(z) = (z-a)^m g(z)$$

where $g(z)$ is holomorphic in an open neighbourhood of a and $g(a) \neq 0$, then

$$\frac{f'(z)}{f(z)}$$

has a simple pole at a with residue m.

Residues of meromorphic functions at poles are closely related to integrals of the functions along paths in $\mathbf{C}$. Recall that a piecewise-smooth path in an open subset W of $\mathbf{C}$ is a continuous map $\gamma : [c, d] \rightarrow W$, defined on a closed interval $[c, d]$ of the real line, which is the *join* of finitely many smooth paths $\gamma_1, \ldots, \gamma_t$ in W. By this we mean that there exist real numbers

$$c = c_0 < c_1 < \ldots < c_t = d$$

and maps $\gamma_i : [c_{i-1}, c_i] \rightarrow W$ for $1 \leq i \leq t$ such that

$$\gamma(t) = \gamma_i(t) \quad \text{if } t \in [c_{i-1}, c_i]$$

and the real and imaginary parts of the restriction of γ_i to the open interval (c_{i-1}, c_i) have derivatives which are continuous and extend continuously to the closed interval $[c_{i-1}, c_i]$. The path γ is *closed* if $\gamma(c) = \gamma(d)$ and *simple* if $\gamma(t) \neq \gamma(s)$ unless $s = t$ or $\{s, t\} = \{c, d\}$.

If $f : W \rightarrow \mathbf{C}$ is a continuous function then the integral of f along a piecewise-smooth path $\gamma : [c, d] \rightarrow W$ is by definition

$$\int_\gamma f(z)dz = \int_c^d f(\gamma(t))\gamma'(t)dt$$

where $\gamma'(t)$ is interpreted as $\gamma_i'(t)$ for $t \in (c_{i-1}, c_i)$.

The fundamental theorem concerning the integrals of holomorphic functions along paths in $\mathbf{C}$ is

Theorem 5.2 (Cauchy's theorem). *Let γ be a contour in $\mathbf{C}$ and let f be a function which is holomorphic inside and on γ. Then*

$$\int_\gamma f(z)dz = 0.$$

Remark 5.3 We shall follow [Priestley 85] and define a *contour* in $\mathbf{C}$ to be a simple closed path in $\mathbf{C}$ which is the join of paths $\gamma_1, \ldots, \gamma_t$, each of which is a straight line segment or circular arc in $\mathbf{C}$. By the Jordan curve theorem (see [Priestley 85] 3.24) the complement in $\mathbf{C}$ of the image γ^* of such a contour γ has two connected components, one bounded (the *inside* $I(\gamma)$ of γ) and one unbounded (the *outside* of γ). The requirement that f be a holomorphic function inside and on γ means that f is holomorphic on some open subset of $\mathbf{C}$ containing $I(\gamma) \cup \gamma^*$. For a proof of this form of Cauchy's theorem we can then refer to [Priestley 85] 4.6. However Cauchy's theorem is true for much more general contours (cf. e.g. exercise 6.2).

A consequence of Cauchy's theorem is

Theorem 5.4 (Cauchy's residue theorem). *Let γ be a contour in $\mathbb{C}$ and let f be a meromorphic function inside and on γ with no poles on γ and poles at $a_1, ..., a_t$ inside γ. Then*

$$\int_\gamma f(z)dz = \pm 2\pi i \sum_{j=1}^{t} res\{f; a_j\}.$$

Proof. [Priestley 85] 7.4. □

Remark 5.5 The sign $\pm$ depends on whether γ is a positively or negatively oriented contour (see [Priestley 85] 4.7). We shall not need the precise definition.

There is a partial converse to Cauchy's theorem.

Theorem 5.6 (Morera's theorem). *If $f : W \to \mathbb{C}$ is a continuous function on an open subset W of $\mathbb{C}$ and if*

$$\int_\gamma f(z)dz = 0$$

for all closed piecewise-smooth paths γ in convex open subsets of W then f is holomorphic in W.

Proof. [Priestley 85] 5.6. □

Using these theorems it is possible to prove the following result which is very useful for constructing holomorphic and meromorphic functions.

Theorem 5.7 *Let $(f_n : W \to \mathbb{C})_{n \geq 1}$ be a sequence of holomorphic functions on an open subset W of $\mathbb{C}$ converging uniformly to a function $f : W \to \mathbb{C}$. Then f is holomorphic on W, and the derivatives f'_n converge uniformly to f' on W.*

Proof. [Priestley 85] p.78. □

A direct application of this result is the Weierstrass M-test.

Theorem 5.8 (Weierstrass' M-test). *Let $(f_n : W \to \mathbb{C})_{n \geq 1}$ be a sequence of holomorphic functions on an open subset W of $\mathbb{C}$. Suppose there exist positive real numbers M_n for $n \geq 1$ such that the series*

$$\sum_{n \geq 1} M_n$$

converges and

$$|f_n(z)| \leq M_n \quad \forall z \in W.$$

Then the series

$$\sum_{n \geq 1} f_n(z)$$

converges uniformly on W to a holomorphic function $f(z)$ such that

$$f'(z) = \sum_{n \geq 1} f_n'(z).$$

Proof. A standard argument shows that the series converges uniformly. The result is then an immediate consequence of theorem 5.7. □.

Remark 5.9 Of course the corresponding result is true if we have a double sequence

$$(f_{n,m} : W \to \mathbf{C})_{n \geq 1, m \geq 1}$$

of holomorphic functions on W.

The most important application of this for our purposes is to define the *Weierstrass $\wp$-function*. Let ω_1 and ω_2 be nonzero complex numbers which are linearly independent over $\mathbf{R}$ (i.e. their quotient ω_1/ω_2 is not real). Let

$$\Lambda = \{n\omega_1 + m\omega_2 \ : \ n, m \in \mathbf{Z}\}.$$

Then Λ is a lattice in $\mathbf{C}$. That is, Λ is an additive subgroup of $\mathbf{C}$ which is isomorphic to $\mathbf{Z} \times \mathbf{Z}$.

Proposition 5.10 *There is a meromorphic function $\wp(z)$ on $\mathbf{C}$ defined by*

$$\wp(z) = z^{-2} + \sum_{\omega \in \Lambda - \{0\}} ((z - \omega)^{-2} - \omega^{-2}),$$

with derivative given by

$$\wp'(z) = \sum_{\omega \in \Lambda} -2(z - \omega)^{-3}.$$

Proof. Note that the sum of a holomorphic function and a meromorphic function on an open subset of $\mathbf{C}$ is a meromorphic function. Hence, by the Weierstrass M-test (theorem 5.8), to show that $\wp(z)$ is a well-defined meromorphic function on $\mathbf{C}$ and that its derivative can be obtained by differentiating the series term by term, it suffices to show that for any $R > 0$ there is a finite subset Λ_R of Λ such that the series

$$\sum_{\omega \in \Lambda - \Lambda_R} ((z - \omega)^{-2} - \omega^{-2})$$

converges absolutely uniformly on the disc

$$\{z \in \mathbf{C} \ : \ |z| \leq R\}.$$

For this we need

Lemma 5.11 *There is some $\delta > 0$ such that*

$$|x\omega_1 + y\omega_2| \geq \delta\sqrt{x^2 + y^2}$$

for all real numbers x and y.

Given this lemma, let

$$\Lambda_R = \{\omega \in \Lambda : |\omega| \leq 2R\}.$$

Then

$$\Lambda_R \subseteq \{n\omega_1 + m\omega_2 : n, m \in \mathbf{Z}, \ n^2 + m^2 \leq 4R^2\delta^{-2}\}$$

and hence Λ_R is finite. Moreover if

$$\omega = n\omega_1 + m\omega_2 \in \Lambda - \Lambda_R$$

and $|z| \leq R$, then $|z| \leq \frac{1}{2}|\omega|$ and hence

$$
\begin{aligned}
|(z - \omega)^{-2} - \omega^{-2}| &= |z(2\omega - z)(z - \omega)^{-2}\omega^{-2}| \\
&\leq (5R|\omega|/2)/(|\omega|^4/4) \\
&= 10R/|\omega|^3 \\
&\leq 10R\delta^{-3}(n^2 + m^2)^{-3/2}.
\end{aligned}
$$

The result now follows by comparing with the series

$$
\begin{aligned}
\sum_{(n,m)\neq(0,0)}(n^2 + m^2)^{-3/2} &= \sum_{k\geq1}\sum_{max(|n|,|m|)=k}(n^2 + m^2)^{-3/2} \\
&\leq 8\sum_{k\geq1}k^{-2} \\
&< \infty.
\end{aligned}
$$

It remains to prove lemma 5.11.

Proof of lemma 5.11. The function $f : [0, 2\pi] \to \mathbf{R}$ defined by

$$f(\theta) = |(\cos\theta)\omega_1 + (\sin\theta)\omega_2|$$

is continuous. Since the interval $[0, 2\pi]$ is compact, f is bounded and attains its bounds. Moreover $f(\theta) > 0$ for all $\theta \in [0, 2\pi]$ since ω_1 and ω_2 are linearly independent over $\mathbf{R}$. Therefore there is some $\delta > 0$ such that

$$f(\theta) > \delta$$

for all $\theta \in [0, 2\pi]$. It follows that

$$|x\omega_1 + y\omega_2| \geq \delta\sqrt{x^2 + y^2}$$

for all $(x, y) \in \mathbf{R} \times \mathbf{R}$. □

Definition 5.12 $\wp(z)$ *is called the Weierstrass ℘-function associated to the lattice* Λ.

Lemma 5.13 $\wp(-z) = \wp(z) = \wp(z + \zeta)$ *for all z in* **C** *and* ζ *in* Λ.

Proof. First note that if $\zeta \in \Lambda$ then

$$\wp'(z + \zeta) = -2 \sum_{\omega \in \Lambda} (z + \zeta - \omega)^{-3}.$$

Since the tail end of this series converges absolutely and since $\omega - \zeta$ runs over Λ as ω runs over Λ, we can rearrange the series and substitute ω for $\omega - \zeta$ to get

$$\wp'(z + \zeta) = \wp'(z)$$

for all $z \in$ **C**. This implies that

$$\wp(z + \zeta) = \wp(z) + c(\zeta)$$

where $c(\zeta)$ depends on ζ but not on z. Substituting $z = -\frac{1}{2}\zeta$ we get

$$c(\zeta) = \wp(\frac{1}{2}\zeta) - \wp(-\frac{1}{2}\zeta).$$

Now observe that

$$\wp(-z) = z^{-2} + \sum_{\omega \in \Lambda - \{0\}} ((z + \omega)^{-2} - \omega^{-2})$$

and we can rearrange this series, replacing ω by $-\omega$, to get

$$\wp(-z) = \wp(z) \quad \forall z \in \mathbf{C}.$$

In particular $c(\zeta) = \wp(\frac{1}{2}\zeta) - \wp(-\frac{1}{2}\zeta) = 0$, and thus

$$\wp(z + \zeta) = \wp(z) \quad \forall z \in \mathbf{C}.$$

This completes the proof. $\square$

Definition 5.14 *Functions f on* **C** *with the property that*

$$f(z + \zeta) = f(z) \quad \forall z \in \mathbf{C}, \forall \zeta \in \Lambda,$$

or equivalently

$$f(z + \omega_1) = f(z) = f(z + \omega_2) \quad \forall z \in \mathbf{C},$$

are called doubly periodic *with period lattice* Λ *(or with periods* ω_1 *and* ω_2*). Thus the Weierstrass ℘-function is a doubly periodic meromorphic function on* **C**.

Lemma 5.15 *A doubly periodic holomorphic function f on* **C** *is constant.*

Proof. Since f is holomorphic it is continuous, and hence by proposition 2.12 is bounded on any closed bounded subset of **C** such as the parallelogram

$$P = \{s\omega_1 + t\omega_2 \; : \; s, t \in [0,1]\}.$$

But given $z \in$ **C** we can find $\zeta \in \Lambda$ such that $z + \zeta \in P$, and since f is doubly periodic,

$$f(z + \zeta) = f(z).$$

Thus f is bounded on **C**, so the result follows from

Theorem 5.16 (Liouville's theorem). *Any bounded holomorphic function on* **C** *is constant.*

Proof. [Priestley 85] 5.2. □

Using lemma 5.15 we can prove an extremely important identity concerning the Weierstrass $\wp$-function.

Lemma 5.17

$$\wp'(z)^2 = 4\wp(z)^3 - g_2\wp(z) - g_3,$$

where

$$g_2 = g_2(\Lambda) = 60 \sum_{\omega \in \Lambda - \{0\}} \omega^{-4}$$

and

$$g_3 = g_3(\Lambda) = 140 \sum_{\omega \in \Lambda - \{0\}} \omega^{-6}.$$

Proof.

$$\wp(z) - z^{-2} = \sum_{\omega \in \Lambda - \{0\}} ((z - \omega)^{-2} - \omega^{-2})$$

vanishes at 0 and restricts to a holomorphic function in an open neighbourhood of 0 in **C**. Moreover it is an even function of z (i.e. it takes the same values at z and $-z$) so its odd derivatives vanish at 0 and hence its Taylor series expansion about 0 involves only even powers of z. Thus near 0 we can write

$$\wp(z) = z^{-2} + \lambda z^2 + \mu z^4 + z^6 h(z)$$

where $h(z)$ is a holomorphic function of z defined near $z = 0$. Then

$$\wp'(z) = -2z^{-3} + 2\lambda z + 4\mu z^3 + 6z^5 h(z) + z^6 h'(z).$$

Consider the function

$$k(z) = \wp'(z)^2 - 4\wp(z)^3 + g_2\wp(z) + g_3$$

where $g_2 = 20\lambda$ and $g_3 = 28\mu$. Using the above formulas for $\wp(z)$ and $\wp'(z)$ we can easily check that $k(z)$ restricts to a holomorphic function of z in an open neighbourhood of 0 which vanishes at 0. But $\wp$ and $\wp'$ are both meromorphic functions on $\mathbf{C}$ which are holomorphic on $\mathbf{C} - \Lambda$ and satisfy

$$\wp(z + \zeta) = \wp(z), \quad \wp'(z + \zeta) = \wp'(z)$$

for all $z \in \mathbf{C}$ and $\zeta \in \Lambda$. This periodicity shows that $k(z)$ is holomorphic in a neighbourhood of each $\zeta \in \Lambda$, and hence $k(z)$ is a doubly periodic holomorphic function on $\mathbf{C}$. Therefore by lemma 5.15, $k(z)$ must be a constant function, so

$$k(z) = k(0) = 0$$

for all $z \in \mathbf{C}$. Finally to obtain the required series expressions for $g_2 = 20\lambda$ and $g_3 = 28\mu$ note that 2λ and 24μ are the second and fourth derivatives at 0 of the function

$$\sum_{\omega \in \Lambda - \{0\}} ((z - \omega)^{-2} - \omega^{-2})$$

which can be differentiated term by term at $z = 0$. □

Proposition 5.18 *The Weierstrass ℘-function*

$$\wp : \mathbf{C} - \Lambda \to \mathbf{C}$$

is surjective. Also

$$\wp(z) = \wp(w)$$

if and only if $w \in \Lambda \pm z$.

Proof. Take any $c \in \mathbf{C}$ and let $f(z) = \wp(z) - c$. Then by remark 5.1 the function $f'(z)/f(z)$ is meromorphic on $\mathbf{C}$ with simple poles of residue $-m$ where f has poles of multiplicity m and simple poles of residue m where f has zeros of multiplicity m, and no other poles. Therefore by Cauchy's residue theorem (theorem 5.4) if γ is a contour in $\mathbf{C}$ not passing through any zero or pole of f then

$$\frac{1}{2\pi i} \int_\gamma \frac{f'(z)}{f(z)} dz = Z - P$$

where Z and P are the numbers of zeros and poles of f inside γ counted according to multiplicity. Take γ to be the boundary of the parallelogram

$$P(a) = \{a + s\omega_1 + t\omega_2 \ : \ s, t \in [0, 1]\}$$

where a is chosen so that the boundary of $P(a)$ does not pass through any zeros or poles of f. In particular this means that there is precisely one lattice point $\zeta \in \Lambda$ inside γ. By the double periodicity of f we have

$$\int_\gamma \frac{f'(z)}{f(z)} dz = 0$$

since the integrals along opposite sides of the parallelogram cancel. Therefore

$$Z = P.$$

But f, like $\wp$, has poles of multiplicity two at all the lattice points $\zeta \in \Lambda$ and no others, so

$$P = 2.$$

Thus $Z = 2$, and hence there exists some $w_0 \in P(a)$ such that

$$f(w_0) = 0$$

that is,

$$\wp(w_0) = c.$$

Since c was arbitrary this shows that $\wp : \mathbf{C} - \Lambda \to \mathbf{C}$ is surjective.

Since $\wp(z)$ is even and doubly periodic we have

$$\wp(z) = \wp(w_0) = c \quad \forall z \in \Lambda \pm w_0.$$

There exists some $w_1 \in \Lambda - w_0$ belonging to the parallelogram $P(a)$, and then both w_0 and w_1 are zeros of f inside γ. So if $w_0 \neq w_1$ these account for the two zeros of f inside γ and hence the only zeros of f are given by

$$z \in \Lambda \pm w_0.$$

Thus it remains to show that if $w_1 = w_0$ then f has a zero of multiplicity at least two at w_0; i.e. that

$$f'(w_0) = 0.$$

For then as before all the zeros of f inside γ will be accounted for. But if $w_0 = w_1$ then $\Lambda + w_0 = \Lambda - w_0$ and so as $\wp'(z)$ is an odd doubly periodic function

$$\wp'(w_0) = -\wp'(-w_0) = -\wp'(w_0).$$

Hence $\wp'(w_0) = 0$ and thus $f'(w_0) = 0$ as required. $\square$

Definition 5.19 *Let C_Λ be the projective curve in $\mathbf{P}_2$ defined by the polynomial*

$$Q_\Lambda(x, y, z) = y^2 z - 4x^3 + g_2 x z^2 + g_3 z^3$$

where $g_2 = g_2(\Lambda)$ and $g_3 = g_3(\Lambda)$ are defined as in lemma 5.17.

Lemma 5.20 *The cubic curve C_Λ is nonsingular.*

Proof. Let

$$\alpha = \wp(\tfrac{1}{2}\omega_1), \ \beta = \wp(\tfrac{1}{2}\omega_2), \ \gamma = \wp(\tfrac{1}{2}(\omega_1 + \omega_2)).$$

To show that C_Λ is nonsingular it is enough to show that α, β, γ are distinct complex numbers and that

$$Q_\Lambda(x, y, z) = y^2 z - 4(x - \alpha z)(x - \beta z)(x - \gamma z).$$

That α, β, γ are distinct follows immediately from proposition 5.18. Since $\wp$ is an even doubly periodic function its derivative is an odd doubly periodic function with the same periods ω_1 and ω_2. Thus as in the proof of (5.18)

$$\wp'(\tfrac{1}{2}\omega_1) = \wp'(\tfrac{1}{2}\omega_1 - \omega_1) = \wp'(-\tfrac{1}{2}\omega_1) = -\wp'(\tfrac{1}{2}\omega_1)$$

and so

$$\wp'(\tfrac{1}{2}\omega_1) = 0.$$

By lemma 5.17 we have

$$4\alpha^3 - g_2\alpha - g_3 = \wp'(\tfrac{1}{2}\omega_1)^2 = 0$$

and so α, and similarly β and γ, are roots of the polynomial $4x^3 - g_2 x - g_3$. It follows that

$$Q_\Lambda(x, y, z) = y^2 z - 4(x - \alpha z)(x - \beta z)(x - \gamma z)$$

with α, β, γ distinct, and hence the curve C_Λ defined by $Q_\Lambda(x, y, z)$ is nonsingular. $\square$

Remark 5.21 If we regard the lattice Λ as an additive subgroup of $\mathbf{C}$, then we can form the quotient group

$$\mathbf{C}/\Lambda = \{\Lambda + a \ : \ a \in \mathbf{C}\}$$

where as usual two cosets $\Lambda + a$ and $\Lambda + b$ are equal if and only if $a - b \in \Lambda$. This quotient group has a natural topology (the quotient topology) induced from the standard topology on $\mathbf{C}$ as follows. Let

$$\pi : \mathbf{C} \to \mathbf{C}/\Lambda$$

be the surjective map defined by $\pi(a) = \Lambda + a$. Then a subset U of $\mathbf{C}/\Lambda$ is open in the quotient topology on $\mathbf{C}/\Lambda$ if and only if its inverse image $\pi^{-1}(U)$ is open in $\mathbf{C}$.

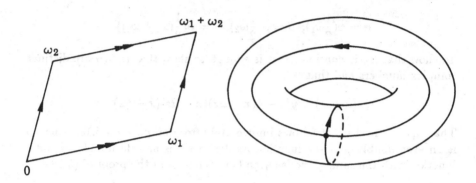

Figure 5.1: $\mathbf{C}/\Lambda$ is a torus

Note that $\pi : \mathbf{C} \to \mathbf{C}/\Lambda$ is an *open map* in the sense that if V is open in $\mathbf{C}$ then $\pi(V)$ is open in $\mathbf{C}/\Lambda$. This is because

$$\pi^{-1}(\pi(V)) = \bigcup_{\zeta \in \Lambda} (V + \zeta)$$

is a union of translates

$$V + \zeta = \{v + \zeta \, : \, v \in V\}$$

of V in $\mathbf{C}$, all of which are open subsets of $\mathbf{C}$.

Note also that $\mathbf{C}/\Lambda$ is compact by 2.12(i) and (ii), since the restriction of π to the parallelogram

$$P = \{s\omega_1 + t\omega_2 \, : \, s, t \in [0, 1]\}$$

is surjective.

Topologically $\mathbf{C}/\Lambda$ is a torus (see figure 5.1).This is because we can identify $\mathbf{C}/\Lambda$ topologically with the parallelogram P with its two pairs of opposite sides glued together. Glueing one pair of sides together gives a cylinder and glueing the ends of the cylinder together gives a torus.

We shall refer to $\mathbf{C}/\Lambda$ as a "complex torus."

Definition 5.22 *The lemmas 5.13 and 5.17 imply that there is a well-defined mapping $u : \mathbf{C}/\Lambda \to \mathbf{C}$ defined by*

$$u(\Lambda + z) = \begin{cases} [\wp(z), \wp'(z), 1] & \text{if } z \notin \Lambda \\ [0, 1, 0] & \text{if } z \in \Lambda. \end{cases}$$

Proposition 5.23 *The mapping $u: \mathbf{C}_\Lambda \to \mathbf{C}$ is a homeomorphism.*

Proof. We need to show that u is a bijection, and that u and u^{-1} are continuous.

First let us check that u is injective. So suppose that $z, w \in \mathbb{C} - \Lambda$ and

$$u(z) = u(w).$$

Then $\wp(z) = \wp(w)$ so by proposition 5.18 we know that

$$z \in \Lambda \pm w.$$

We want to show that $z \in \Lambda + w$. Suppose $z \in \Lambda - w$. Then since $\wp'$ is an odd doubly periodic function we have

$$\wp'(z) = -\wp'(w).$$

But $u(z) = u(w)$ so $\wp'(z) = \wp'(w)$. Hence

$$\wp'(z) = \wp'(w) = 0.$$

But the proof of lemma 5.20 shows that if $\wp'(w) = 0$ then $\wp(w)$ equals α, β or γ where

$$\alpha = \wp(\tfrac{1}{2}\omega_1), \; \beta = \wp(\tfrac{1}{2}\omega_2), \; \gamma = \wp(\tfrac{1}{2}(\omega_1 + \omega_2)),$$

and then (5.18) implies that

$$w \in \frac{1}{2}\Lambda,$$

that is, $\Lambda + w = \Lambda - w$. So $z \in \Lambda + w$ and we have shown that u is injective.

Now let us show that u is surjective. So suppose that $[a, b, c] \in C_\Lambda$. If $c = 0$ then the equation defining C_Λ forces $a = 0$ so $[a, b, c] = [0, 1, 0]$ which is in the image of u. Otherwise we may assume $c = 1$. Then by (5.18) there is some $z \in \mathbb{C}$ such that

$$\wp(z) = a.$$

By lemma 5.17 and the assumption that $[a, b, 1] \in C_\Lambda$ we have

$$\wp'(z)^2 = 4\wp(z)^3 - g_2\wp(z) - g_3 = 4a^3 - g_2 a - g_3 = b^2$$

so $\wp'(z) = \pm b$. Since $\wp$ is an even function and $\wp'$ is an odd function it follows that either

$$u(\Lambda + z) = [a, b, 1]$$

or

$$u(\Lambda - z) = [a, b, 1].$$

Thus u is surjective.

Since $\wp$ and $\wp'$ are holomorphic and hence continuous functions on $\mathbf{C} - \Lambda$ it is clear that u is continuous except possibly at $\Lambda + 0$. Since $\wp$ and $\wp'$ have poles of multiplicity two and three at 0, near 0 we can write

$$\wp(z) = \frac{g(z)}{z^2}$$

$$\wp'(z) = \frac{h(z)}{z^3}$$

where g and h are holomorphic near 0 and $g(0)$ and $h(0)$ are nonzero. So if z is near 0 but not equal to 0 then

$$u(\Lambda + z) = [\wp(z), \wp'(z), 1] = [zg(z), h(z), z^3]$$

which tends to $[0, 1, 0]$ as z tends to 0. Thus u is continuous at $\Lambda + 0$.

We have shown that $u : \mathbf{C}/\Lambda \to C_\Lambda$ is a continuous bijection. Since $\mathbf{C}/\Lambda$ is compact (by remark 5.21) and C_Λ is Hausdorff (lemma 2.30) it follows that u is a homeomorphism ([Sutherland 75]). $\square$

Remark 5.24 We shall show later (proposition 5.43) that u and its inverse are not only continuous but indeed holomorphic. However before proving this we must say what it means! This will be done in the next section.

Note also that proposition 5.23 tallies with the degree-genus formula, which tells us that topologically C_Λ is a sphere with one handle, i.e. a torus.

5.2   Riemann surfaces

We saw in the last section that if

$$\Lambda = \{n\omega_1 + m\omega_2 : n, m \in \mathbf{Z}\}$$

is a lattice in $\mathbf{C}$ then there is a nonsingular cubic curve C_Λ in $\mathbf{P}_2$ and a homeomorphism

$$u : \mathbf{C}/\Lambda \to C_\Lambda$$

defined by

$$u(\Lambda + z) = \begin{cases} [\wp(z), \wp'(z), 1] & \text{if } z \notin \Lambda \\ [0, 1, 0] & \text{if } z \in \Lambda \end{cases}$$

where $\wp(z)$ is the Weierstrass $\wp$-function associated to Λ. In fact this homeomorphism u and its inverse u^{-1} are *holomorphic*. The main aim of this section is to make sense of this statement (which will be proved at the end of it: see proposition 5.43). We need to define Riemann surfaces, show that the complex torus $\mathbf{C}/\Lambda$ and the nonsingular curve C_Λ can each be regarded as a

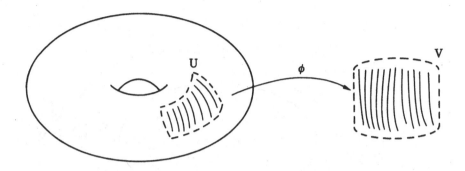

Figure 5.2: A chart

Riemann surface in a natural way and say what it means for a map between Riemann surfaces to be holomorphic.

First we need the definition of a (topological) surface.

Definition 5.25 *A surface is a Hausdorff topological space S which is locally homeomorphic to* **C** *(or equivalently* **R**^{2}*).*

Here, "locally homeomorphic to **C**" means that any $x \in S$ has an open neighbourhood U in S which is homeomorphic to an open subset V of **C**.

A homeomorphism $\phi : U \to V$ between an open subset U of S and an open subset V of **C** is called a *chart* (or local coordinate mapping) on S (see figure 5.2).An *atlas* Φ for the surface S is a collection of charts on S

$$\Phi = \{\phi_\alpha : U_\alpha \to V_\alpha : \alpha \in A\}$$

indexed by a set A, such that

$$S = \bigcup_{\alpha \in A} U_\alpha.$$

If $\phi_\alpha : U_\alpha \to V_\alpha$ and $\phi_\beta : U_\beta \to V_\beta$ are charts then

$$\phi_\alpha(U_\alpha \cap U_\beta)$$

is an open subset of V_α, which is an open subset of **C**. If $\Phi = \{\phi_\alpha : U_\alpha \to V_\alpha : \alpha \in A\}$ is an atlas for S then the homeomorphisms

$$\phi_{\alpha\beta} = \phi_\alpha \circ \phi_\beta^{-1} : \phi_\beta(U_\alpha \cap U_\beta) \to \phi_\alpha(U_\alpha \cap U_\beta)$$

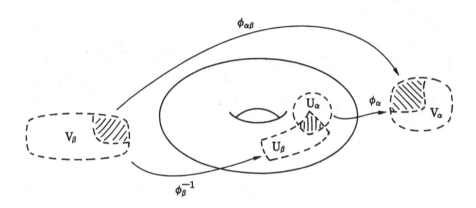

Figure 5.3: A transition function

between open subsets of **C** are called the *transition functions* of the atlas (see figure 5.3). The atlas is called *holomorphic* if all its transition functions are holomorphic in the usual sense as functions from open subsets of **C** to open subsets of **C**.

Examples 5.26 If U is an open subset of **C** then the identity map $\{1_U : U \to U\}$ is a holomorphic atlas on U. So is the set of all holomorphic functions with holomorphic inverses from open subsets of U to open subsets of **C**.

Nonsingular complex curves are the most important examples of surfaces with holomorphic atlases from the point of view of this book.

Proposition 5.27 *If C is a complex algebraic curve in $\mathbf{C}^2$ defined by a polynomial $P(x,y)$ then $C - Sing(C)$ has a holomorphic atlas.*

Proof. Suppose that $(a,b) \in C$, i.e. $P(a,b) = 0$, and that

$$(\partial P/\partial y)(a,b) \neq 0.$$

The implicit function theorem (see Appendix B) tells us that there are open neighbourhoods V and W of a and b in **C** and a holomorphic function g :

$V \to W$ such that if $x \in V$ and $y \in W$ then

$$P(x,y) = 0 \iff y = g(x).$$

Since $C - Sing(C)$ is an open subset of C (this follows immediately from the definition of the quotient topology on $\mathbf{P}_2$; see after definition 2.15), by choosing V and W small enough we may assume that

$$U = \{(x,y) \in C : x \in V, y \in W\}$$

is an open neighbourhood of (a,b) in $C - Sing(C)$. Then the map $\phi : U \to V$ defined by

$$\phi(x,y) = x$$

is a continuous function with continuous inverse

$$x \mapsto (x, g(x)).$$

Similarly if $(\partial P / \partial x)(a,b) \neq 0$ there is an open neighbourhood U of (a,b) in $C - Sing(C)$ such that the map $\psi : U \to \mathbf{C}$ defined by

$$\psi(x,y) = y$$

is a homeomorphism onto an open subset V of $\mathbf{C}$ with inverse

$$y \mapsto (h(y), y)$$

where $h(y)$ is a holomorphic function of y. Thus there is an atlas on $C - Sing(C)$ such that every chart is of one of these two forms ϕ or ψ . The transition functions are then either the identity or compositions of the form

$$x \mapsto (x, g(x)) \mapsto g(x)$$

or

$$y \mapsto (h(y), y) \mapsto h(y)$$

where g and h are holomorphic. Therefore this atlas is holomorphic. $\Box$

The argument for projective curves is very similar.

Proposition 5.28 *If C is a projective curve in $\mathbf{P}_2$ defined by a homogeneous polynomial $P(x,y,z)$ then $C - Sing(C)$ has a holomorphic atlas.*

Proof. Suppose that $[a,b,c] \in C$, i.e. $P(a,b,c) = 0$, and that

$$(\partial P / \partial y)(a,b,c) \neq 0.$$

By Euler's relation (see lemma 2.32)

$$a\frac{\partial P}{\partial x}(a,b,c) + b\frac{\partial P}{\partial y}(a,b,c) + c\frac{\partial P}{\partial z}(a,b,c) = 0$$

so

$$a = c = 0 \Longrightarrow a = b = c = 0$$

which is impossible by the definition of $\mathbf{P_2}$. Hence $a \neq 0$ or $c \neq 0$. Assume that $c \neq 0$. Then by the homogeneity of P

$$\frac{\partial P}{\partial y}(a/c, b/c, 1) = c^{-(d-1)}\frac{\partial P}{\partial y}(a,b,c) \neq 0$$

where d is the degree of P. The implicit function theorem (see Appendix B) applied to the polynomial $P(x,y,1)$ in x and y now tells us that there are open neighbourhoods V and W of a/c and b/c in $\mathbf{C}$ and a holomorphic function $g : V \to W$ such that if $x \in V$ and $y \in W$ then

$$P(x,y,1) = 0 \iff y = g(x).$$

If V and W are chosen small enough then

$$
\begin{aligned}
U &= \{[x,y,z] \in C : z \neq 0, x/z \in V, y/z \in W\} \\
&= \{[x,y,1] \in C : x \in V, y \in W\}
\end{aligned}
$$

is an open neighbourhood of $[a,b,c]$ in $C - Sing(C)$. The map $\phi : U \to V$ defined by

$$\phi[x,y,z] = x/z$$

is a homeomorphism with inverse

$$w \mapsto [w, g(w), 1].$$

Similarly if $[a,b,c] \in C$ and $(\partial P/\partial y)(a,b,c) \neq 0 \neq a$, or if $(\partial P/\partial x)(a,b,c) \neq 0$ or $(\partial P/\partial z)(a,b,c) \neq 0$, then we can find a homeomorphism $\phi : U \to V$ from an open neighbourhood U of $[a,b,c]$ in $C - Sing(C)$ onto an open subset V of $\mathbf{C}$ such that $\phi[x,y,z]$ is one of the following:

$$z/x, y/z, z/y, x/y, y/x$$

and the inverse of ϕ has the form

$$w \mapsto [1, g(w), w], [g(w), w, 1], [g(w), 1, w], [w, 1, g(w)] \text{ or } [1, w, g(w)]$$

where $g : V \to \mathbf{C}$ is holomorphic. Thus we get an atlas on $C - Sing(C)$ each of whose transition functions is of one of the following forms

$$w \mapsto w, 1/w, g(w), 1/g(w), w/g(w) \text{ or } g(w)/w$$

such that g is holomorphic and the denominator does not vanish on the set where the transition function is defined. Thus this atlas is holomorphic. $\square$

Remark 5.29 It is easy to show that the complement of a finite set of points in a surface is dense. Thus the claim in the proof of proposition 4.22 that the complement of a finite set of points in a nonsingular projective curve is dense is substantiated by proposition 5.28.

The point of having a holomorphic atlas on a surface is that it makes it possible to talk about holomorphic functions on the surface.

Definition 5.30 *Let* $\Phi = \{\phi_\alpha : U_\alpha \to V_\alpha : \alpha \in A\}$ *be a holomorphic atlas on a surface* S. *A continuous map* $f : S \to \mathbf{C}$ *is called* holomorphic *with respect to* Φ *at* $x \in S$ *if there is a chart* $\phi_\alpha : U_\alpha \to V_\alpha$ *in* Φ *such that* $x \in U_\alpha$ *and*

$$f \circ \phi_\alpha^{-1} : V_\alpha \to \mathbf{C}$$

is holomorphic at $\phi_\alpha(x)$ *in the usual sense as a function from the open subset* V_α *of* $\mathbf{C}$ *to* $\mathbf{C}$. *The map* f *is called* holomorphic *with respect to* Φ *if it is holomorphic at every* $x \in S$.

Remark 5.31 Because Φ is a holomorphic atlas the choice of chart $\phi_\alpha :$ $U_\alpha \to V_\alpha$ such that $x \in U_\alpha$ in this definition makes no difference. If $x \in U_\alpha \cap U_\beta$ and

$$\phi_\alpha : U_\alpha \to V_\alpha, \phi_\beta : U_\beta \to V_\beta$$

are charts in Φ then $\phi_\alpha(U_\alpha \cap U_\beta)$ is an open subset of V_α containing $\phi_\alpha(x)$ and

$$f \circ \phi_\alpha^{-1}\,|_{\phi_\alpha(U_\alpha \cap U_\beta)} = (f \circ \phi_\beta^{-1}) \circ (\phi_\beta \circ \phi_\alpha^{-1})\,|_{\phi_\alpha(U_\alpha \cap U_\beta)}\,.$$

Since $\phi_\beta \circ \phi_\alpha^{-1}$ and its inverse $\phi_\alpha \circ \phi_\beta^{-1}$ are both holomorphic and the composition of holomorphic functions is always holomorphic, it follows that $f \circ \phi_\alpha^{-1}$ is holomorphic at $\phi_\alpha(x)$ if and only if $f \circ \phi_\beta^{-1}$ is holomorphic at $\phi_\beta(x)$.

This remark proves

Lemma 5.32 *A continuous function* $f : S \to \mathbf{C}$ *is holomorphic with respect to a holomorphic atlas* Φ *on* S *if and only if*

$$f \circ \phi_\alpha^{-1} : V_\alpha \to \mathbf{C}$$

is holomorphic for every chart $\phi_\alpha : U_\alpha \to V_\alpha$ *in* Φ.

Definition 5.33 *Let* S *and* T *be surfaces with holomorphic atlases* Φ *and* Ψ. *A continuous map* $f : S \to T$ *is called* holomorphic *with respect to* Φ *and* Ψ *if*

$$\psi_\beta \circ f \circ \phi_\alpha^{-1}\,|_{\phi_\alpha(U_\alpha \cap f^{-1}(W_\beta))} : \phi_\alpha(U_\alpha \cap f^{-1}(W_\beta)) \longrightarrow Y_\beta$$

is holomorphic for every chart $\phi_\alpha : U_\alpha \to V_\alpha$ *in* Φ *and every chart* $\psi_\beta : W_\beta \to Y_\beta$ *in* Ψ.

Note that $f^{-1}(W_\beta)$ is an open subset of S because f is continuous, and thus $\phi_\alpha(U_\alpha \cap f^{-1}(W_\beta))$ is an open subset of $\mathbb{C}$. We can replace the condition that f should be continuous by the weaker condition that $f^{-1}(W_\beta)$ should be an open subset of S for all β, or equivalently that $\phi_\alpha(U_\alpha \cap f^{-1}(W_\beta))$ should be an open subset of $\mathbb{C}$ for all α and β.

Lemma 5.34 *If $f : S \to T$ and $g : T \to R$ are holomorphic with respect to given holomorphic atlases Φ, Ψ, Θ on the surfaces S, T, R then $g \circ f : S \to R$ is holomorphic with respect to the holomorphic atlases Φ, Θ on S and R.*

Proof. Suppose $x \in S$; we need to show that $g \circ f$ is holomorphic at x. Let us choose holomorphic charts

$$\phi_\alpha : U_\alpha \to V_\alpha \subseteq \mathbb{C}$$

on S,

$$\psi_\beta : W_\beta \to X_\beta \subseteq \mathbb{C}$$

on T and

$$\theta_\gamma : Y_\gamma \to Z_\gamma \subseteq \mathbb{C}$$

on R such that

$$x \in U_\alpha, \; f(x) \in W_\beta, \; g(f(x)) \in Y_\gamma.$$

It is enough to show that $\theta_\gamma \circ g \circ f \circ \phi_\alpha^{-1}$ is holomorphic at $\phi_\alpha(x) \in V_\alpha$. But in the open neighbourhood

$$\phi_\alpha(U_\alpha \cap f^{-1}(W_\beta) \cap f^{-1} g^{-1}(Y_\gamma))$$

of $\phi_\alpha(x)$ we can write

$$\theta_\gamma \circ g \circ f \circ \phi_\alpha^{-1} = (\theta_\gamma \circ g \circ \psi_\beta^{-1}) \circ (\psi_\beta \circ f \circ \phi_\alpha^{-1}$$

(see figure 5.4) and this is the composition of two holomorphic functions in the usual sense of complex analysis, so it is holomorphic. $\square$

Definition 5.35 *Two holomorphic atlases Φ and Ψ on a surface S are called compatible if the identity map $1_S : S \to S$ is holomorphic both as a map from S with atlas Φ to S with atlas Ψ and as a map from S with atlas Ψ to S with atlas Φ.*

It is easy to check that compatibility is an equivalence relation on the set of holomorphic atlases on a surface S.

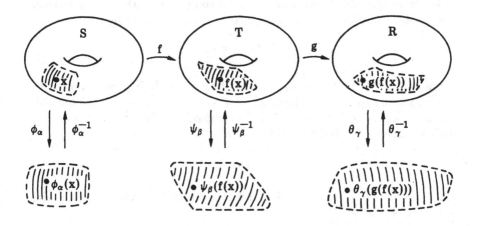

Figure 5.4: Composition of holomorphic functions

Example 5.36 The holomorphic atlases

$$\{1_S : S \to S\}$$

and·

$$\{h : U \to V \ : \ U, V \ \text{open in } \mathbf{C}, \ h \ \text{holomorphic with holomorphic inverse} \ \}$$

on $\mathbf{C}$ are compatible. These atlases are not compatible with the holomorphic atlas on $\mathbf{C}$ consisting of the single chart $\phi : \mathbf{C} \to \mathbf{C}$ defined by

$$\phi(z) = \overline{z}.$$

Remark 5.37 It follows from lemma 5.43 by considering the composition $1_T \circ f \circ 1_S$ with appropriate atlases that if Φ and Ψ are compatible holomorphic atlases on a surface S and $\tilde{\Phi}$ and $\tilde{\Psi}$ are compatible holomorphic atlases on a surface T then a continuous function

$$f : S \to T$$

is holomorphic with respect to Φ and $\tilde{\Phi}$ if and only if it is holomorphic with respect to Ψ and $\tilde{\Psi}$.

Definition 5.38 *A Riemann surface is a surface S together with an equivalence class $\mathcal{H}$ of holomorphic atlases on S. In other words a Riemann surface is given by a surface S with a holomorphic atlas on S, and two holomorphic atlases on S define the same Riemann surface if and only if they are compatible with each other.*

If $(S, \mathcal{H})$ and $(T, \mathcal{F})$ are Riemann surfaces and $f : S \to T$ is continuous then by remark 5.37 we can say that f is a *holomorphic* map between the Riemann surfaces $(S, \mathcal{H})$ and $(T, \mathcal{F})$ if it is holomorphic with respect to any (or equivalently all) of the holomorphic atlases $\Phi \in \mathcal{H}$ on S and $\Psi \in \mathcal{F}$ on T.

We often denote a Riemann surface $(S, \mathcal{H})$ by the same symbol S as the underlying surface, unless confusion is likely to arise.

Definition 5.39 *Two Riemann surfaces S and T are called* biholomorphic *if there is a holomorphic bijection $f : S \to T$ whose inverse is holomorphic. (In fact the requirement that the inverse of f should be holomorphic is redundant: see exercise 5.6).*

Examples 5.40

(a). Any open subset U of $\mathbf{C}$ with the holomorphic atlas $\{1_U : U \to U\}$ is a Riemann surface.

(b). If S is a Riemann surface with holomorphic atlas

$$\Phi = \{\phi_\alpha : U_\alpha \to V_\alpha \; : \; \alpha \in A\}$$

and if W is an open subset of S then the holomorphic atlas

$$\Phi \mid_W = \{\phi_\alpha \mid_{U_\alpha \cap W} : U_\alpha \cap W \to \phi_\alpha(U_\alpha \cap W) \; : \; \alpha \in A\}$$

makes W into a Riemann surface.

Note in particular that it is easy to prove from definition 5.25 that every connected component of a Riemann surface S is an open subset of S and hence is a Riemann surface.

(c). Proposition 5.27 tells us that every nonsingular complex curve C in $\mathbf{C}^2$ can be regarded as a Riemann surface in a natural way. The restrictions to C of the x and y-coordinates on $\mathbf{C}^2$ are then holomorphic functions on C, as are all polynomial functions of x and y. Similarly it follows from propositions 5.27 and 5.28 that if C is an affine or projective complex curve then $C - Sing(C)$ is a Riemann surface.

(d). The *Riemann sphere* is $\mathbf{P}_1 = \mathbf{C} \cup \{\infty\}$ (sometimes written $\hat{\mathbf{C}}$) where $z \in \mathbf{C}$ is identified with $[z,1] \in \mathbf{P}_1$ and ∞ is $[1,0]$.

Let $U = \mathbf{P}_1 - \{\infty\}$ and let $V = \mathbf{P}_1 - \{0\}$. Define $\phi : U \to \mathbf{C}$ and $\psi : V \to \mathbf{C}$ by

$$\phi[x,y] = x/y, \quad \psi[x,y] = y/x.$$

Then $\phi : U \to \mathbf{C}$ and $\psi : V \to \mathbf{C}$ form a holomorphic atlas on $\mathbf{P}_1$ with transition functions

$$\phi \circ \psi^{-1} = \psi \circ \phi^{-1} : \mathbf{C} - \{0\} \to \mathbf{C} - \{0\}$$

both given by

$$z \mapsto 1/z.$$

Recall that if U is an open subset of $\mathbf{C}$ then a *meromorphic* function on U is a function $f : U \to \mathbf{C} \cup \{\infty\}$ such that if $a \in U$ there exists $\varepsilon > 0$ such that in the punctured disc

$$\{z \in \mathbf{C} : 0 < \mid z - a \mid < \varepsilon\}$$

the function f takes only finite values and has a Laurent series expansion

$$f(z) = \sum_{k \geq -m} c_k (z - a)^k$$

for some integer m. If this power series is identically zero then $f(a) = 0$. Otherwise we may assume that $c_{-m} \neq 0$, and then

$$f(a) = \begin{cases} \infty & \text{if } m > 0, \\ c_0 & \text{if } m = 0, \\ 0 & \text{if } m < 0. \end{cases}$$

When $f(a) = \infty$ we say that f has a *pole* of order m at a.

Equivalently we can write

$$f(z) = (z - a)^{-m} h(z)$$

where

$$h(z) = \sum_{k \geq 0} c_{k-m}(z - a)^k$$

is holomorphic and $h(a) \neq 0$. Thus $f(z)$ is meromorphic in a neighbourhood of a with a pole at a if and only if

$$1/f(z) = (z - a)^m / h(z)$$

is holomorphic near a with value 0 at a. This means that a meromorphic function on U is holomorphic in the sense of Riemann surfaces as a mapping from U to $\mathbf{P}_1 = \mathbf{C} \cup \{\infty\}$. Conversely any holomorphic function $f : U \to \mathbf{P}_1$ which does not take the constant value ∞ on any connected component of U defines a meromorphic function on U.

If

$$p(z) = a_0 + a_1 z + \dots + a_n z^n$$

and

$$q(z) = b_0 + b_1 z + \dots + b_m z^m$$

are polynomials of degrees n and m with no common factor then there is a holomorphic map $f : \mathbf{P}_1 \to \mathbf{P}_1$ whose restriction to $\mathbf{C}$ is the rational function

$$p(z)/q(z)$$

and whose value at ∞ is

$$f(\infty) = \begin{cases} 0 & \text{if } m > n, \\ a_n/b_n & \text{if } m = n \\ \infty & \text{if } m < n. \end{cases}$$

Note that unless f is constant it must take the value ∞ at some point. In fact every nonconstant holomorphic function $f : \mathbf{P}_1 \to \mathbf{P}_1$ is of this form.

Lemma 5.41 *Any nonconstant holomorphic map $f : \mathbf{P}_1 \to \mathbf{P}_1$ is rational (and hence takes the value ∞ at least once).*

Proof. The restriction of f to $\mathbf{C}$ is a meromorphic function on $\mathbf{C}$, as is the function

$$z \mapsto f(\frac{1}{z})$$

where we interpret $\frac{1}{0}$ as ∞. If a meromorphic function has a pole at a point $a \in \mathbf{C}$ then it is holomorphic on some punctured disc

$$\{z \in \mathbf{C} : 0 <| z - a |< \varepsilon\}$$

with $\varepsilon > 0$ (this follows straight from the definition). Since $\mathbf{P}_1$ is compact it follows that f can have only finitely many poles in $\mathbf{C}$, at $a_1, \ldots, a_k$ say. Suppose that

$$f(z) = \sum_{n \geq -m_j} c_n^{(j)}(z - a_j)^n$$

is the Laurent expansion of f about a_j, and let

$$g(z) = \sum_{j=1}^{k} \sum_{n=-m_j}^{-1} c_n^{(j)}(z - a_j)^n.$$

Then g is a rational, and hence meromorphic function on $\mathbf{C}$ which extends to a holomorphic function $f : \mathbf{P}_1 \to \mathbf{P}_1$ such that $g(\infty) = 0$. Moreover $f - g$ has no poles in $\mathbf{C}$, so it has a Taylor expansion

$$f(z) - g(z) = \sum_{n \geq 0} c_n z^n$$

valid for all $z \in \mathbf{C}$. Therefore the Laurent expansion of $f - g$ about ∞ with respect to the local chart given by $w = \frac{1}{z}$ is

$$\sum_{n \geq 0} c_n w^{-n}$$

This expression only defines a meromorphic function if $c_n = 0$ for all sufficiently large n (otherwise there is an essential singularity: see e.g. [Priestley 85] 6.6). Thus $f - g$ is a polynomial on $\mathbf{C}$. Hence f is the sum of two rational functions and so it is itself rational. $\square$

We have one last important example of Riemann surfaces.

Example 5.42 Complex tori. As in §5.1 let ω_1 and ω_2 be complex numbers which are linearly independent over $\mathbf{R}$. Then the lattice

$$\Lambda = \{n\omega_1 + m\omega_2 : n, m \in \mathbf{Z}\}$$

in C is an additive subgroup of C. We have already noted in remark 5.21 that the quotient group C/Λ has a natural topology, the quotient topology induced by the map $\pi : C \to C/\Lambda$ which maps

$$z \mapsto \Lambda + z.$$

It follows immediately from lemma 5.11 that there exists $\delta > 0$ such that

$$\mid n\omega_1 + m\omega_2 \mid \geq \delta$$

for all pairs of integers n, m not both zero. Hence if $a \in C$ the restriction of π to the open disc

$$U_a = \{z \in C : |z - a| < \frac{1}{4}\delta\}$$

in C is a homeomorphism onto the open subset $\pi(U_a)$ of C/Λ. If $\pi(U_a) \cap \pi(U_b) \neq \emptyset$ then there is a *unique*

$$n\omega_1 + m\omega_2 \in \Lambda$$

such that

$$\mid n\omega_1 + m\omega_2 + a - b \mid < \frac{1}{2}\delta,$$

and then

$$(\pi \mid_{U_b})^{-1} \circ \pi \mid_{U_a \cap \pi^{-1}(\pi(U_b))} : U_a \cap \pi^{-1}(\pi(U_b)) \to U_b$$

is given by translation by $n\omega_1 + m\omega_2$. Therefore the charts

$$\phi_a = (\pi \mid_{U_a})^{-1} : \pi(U_a) \to U_a$$

for $a \in C$ form a holomorphic atlas on C/Λ.

To show that this atlas makes C/Λ into a Riemann surface it remains to show that C/Λ is Hausdorff. This can be checked directly using lemma 5.11 but it also follows immediately from the existence of a homeomorphism $u : C/\Lambda \to C_\Lambda$ where C_Λ is a nonsingular projective curve (proposition 5.23) and the fact that C_Λ is Hausdorff (lemma 2.30).

If S is another Riemann surface then a function f from C/Λ to S is holomorphic if and only if the composition $f \circ \pi$ from C to S is holomorphic. Thus holomorphic functions f from C/Λ to S correspond exactly to holomorphic functions g from C to S with the property that

$$g(z + \omega_1) = g(z) = g(z + \omega_2)$$

for all $z \in C$. Functions with this property are called *doubly periodic* with periods ω_1 and ω_2.

Finally we can prove as promised that the homeomorphism $u : C/\Lambda \to C_\Lambda$ and its inverse are holomorphic.

Proposition 5.43 *The homeomorphism* $u : \mathbf{C}/\Lambda \to C_\Lambda$ *defined by*

$$u(\Lambda + z) = \begin{cases} [\wp(z), \wp'(z), 1] & \text{if } z \notin \Lambda \\ [0, 1, 0] & \text{if } z \in \Lambda \end{cases}$$

is holomorphic. So is its inverse $u^{-1} : C_\Lambda \to \mathbf{C}/\Lambda$.

Proof. To show that u is holomorphic we must check that if $\Lambda + w \in \mathbf{C}/\Lambda$ then

$$\psi_\beta \circ u \circ \phi_\alpha^{-1} : \phi_\alpha(U_\alpha \cap u^{-1}(W_\beta)) \to Y_\beta$$

is holomorphic in the usual sense of complex analysis for some holomorphic charts $\phi_\alpha : U_\alpha \to V_\alpha$ on $\mathbf{C}/\Lambda$ and $\psi_\beta : W_\beta \to Y_\beta$ on C_Λ such that $\Lambda + w \in U_\alpha$ and $u(\Lambda + w) \in W_\beta$. As in example 5.42 we may take ϕ_α to be the inverse of

$$\pi : V_\alpha \to U_\alpha = \pi(V_\alpha)$$

where V_α is a sufficiently small open disc in $\mathbf{C}$ and $\pi : \mathbf{C} \to \mathbf{C}/\Lambda$ maps z to its coset $\Lambda + z$. Thus $\phi_\alpha^{-1} = \pi : V_\alpha \to U_\alpha$.

If $w \notin \Lambda$ then $u(\Lambda + w) = [\wp(z), \wp'(z), 1]$ so by the proof of proposition 5.28 we can take ψ_β to be either

$$[x, y, z] \mapsto \frac{x}{z}$$

or

$$[x, y, z] \mapsto \frac{y}{z}.$$

Thus $\psi_\beta \circ u \circ \phi_\alpha^{-1}$ is the restriction of either $\wp$ or $\wp'$, both of which are holomorphic near w.

If $w \in \Lambda$ so that $u(\Lambda + w) = [0, 1, 0]$ the proof of proposition 5.28 shows that we can take ψ_β to be the map

$$[x, y, z] \mapsto \frac{x}{y}$$

since

$$\frac{\partial Q_\Lambda}{\partial z}(0, 1, 0) \neq 0$$

(see definition 5.19). Thus

$$\psi_\beta \circ u \circ \phi_\alpha^{-1}(z) = \begin{cases} \wp(z)/\wp'(z) & \text{if } z \notin \Lambda \\ 0 & \text{if } z \in \Lambda. \end{cases}$$

This function is holomorphic near 0 (and hence near every point of Λ by double periodicity) since near 0 we can write

$$\wp(z) = \frac{g(z)}{z^2}$$

and

$$\wp'(z) = \frac{h(z)}{z^3}$$

where g and h are holomorphic near 0 and $g(0)$ and $h(0)$ are nonzero (cf. the proof of proposition 5.23).

This completes the proof that u is holomorphic. To show that u^{-1} is holomorphic we must check that

$$\phi_\alpha \circ u^{-1} \circ \psi_\beta^{-1} = (\psi_\beta \circ u \circ \phi_\alpha^{-1})^{-1}$$

is holomorphic for every holomorphic chart $\phi_\alpha : U_\alpha \to V_\alpha$ on $\mathbf{C}/\Lambda$ and $\psi_\beta : W_\beta \to Y_\beta$ on C_Λ. This can be proved directly as was done for u, but it also follows immediately from the inverse function theorem of complex analysis ([Priestley 85] 10.25) which tells us that if $f : U \to V$ is a holomorphic bijection between open subsets of $\mathbf{C}$ then $f^{-1} : V \to U$ is also holomorphic. □

5.3   Exercises

5.1. The identity theorem (or principle of analytic continuation) of complex analysis tells us that if $f : U \to \mathbf{C}$ and $g : U \to \mathbf{C}$ are holomorphic functions from a connected open subset U of $\mathbf{C}$ to $\mathbf{C}$, and if $f(z) = g(z)$ for all z in some nonempty open subset W of U then $f(z) = g(z)$ for all $z \in U$ (cf. [Priestley 85] 5.16). Use this to show that if $f : R \to S$ and $g : R \to S$ are holomorphic functions between Riemann surfaces R and S and if R is connected and $f(z) = g(z)$ for all z in some nonempty open subset W of R then $f(z) = g(z)$ for all $z \in R$. [Hint: show that the union of all the open subsets of R on which f and g agree is nonempty and is both open and closed in R, so it must be R itself].

5.2. The open mapping theorem of complex analysis (see [Priestley 85] 10.24) says that if $f : U \to \mathbf{C}$ is a nonconstant holomorphic function from a connected open subset U of $\mathbf{C}$ to $\mathbf{C}$ then $f(U)$ is open in $\mathbf{C}$. Use this to show that if $f : R \to S$ is a nonconstant holomorphic map between Riemann surfaces and R is connected then $f(R)$ is an open subset of S.

5.3. Show that a nonconstant holomorphic function $f : R \to S$ between connected Riemann surfaces R and S such that R is compact is surjective. [Hint: use exercise 5.2 and the fact that the image of a compact space under a continuous map to a Hausdorff space is compact and hence closed]. Deduce that S is compact.

5.4. Show that if R is a compact connected Riemann surface then there are no nonconstant holomorphic functions $f : R \to \mathbf{C}$.

[Hint: Use exercise 5.3]

5.5. Use the theorem of isolated zeros from complex analysis (see [Priestley 85] 5.14) to show that if $f : R \to S$ is a nonconstant holomorphic map between connected Riemann surfaces then every $x \in R$ has an open neighbourhood U in R such that $f(y) \neq f(x)$ for all $y \in U - \{x\}$.

5.6. The inverse function theorem of complex analysis (see [Priestley 85] 10.25) says that if $f : U \to V$ is a holomorphic bijection between open subsets of $\mathbf{C}$ then $f'(z) \neq 0$ for all $z \in U$ and $f^{-1} : V \to U$ is holomorphic. Deduce that if $f : R \to S$ is a holomorphic bijection between Riemann surfaces then $f^{-1} : S \to R$ is holomorphic.

5.7. (a). Show that the union of two compatible holomorphic atlases on a surface S is a holomorphic atlas.
(b). A holomorphic atlas on a surface S is called complete if it contains every holomorphic atlas which is compatible with it. Show that in every equivalence class $\mathcal{H}$ of compatible holomorphic atlases on S there is a unique complete holomorphic atlas.

5.8. Let S be a surface with open subsets V and W such that $s = V \cup W$. Suppose that V and W have holomorphic atlases Φ and Ψ such that the holomorphic atlases Φ $|_{V \cap W}$ and Ψ $|_{V \cap W}$ on $V \cap W$ are compatible. Show that $\Phi \cup \Psi$ is a holomorphic atlas on S.

5.9. Use exercise 5.4 and example 5.42 to show that there are no nonconstant holomorphic doubly periodic functions $f : \mathbf{C} \to \mathbf{C}$.

5.10. Let $\wp(z)$ be the Weierstrass $\wp$-function associated with the lattice Λ in $\mathbf{C}$. Explain why the function

$$\tilde{\wp} : \mathbf{C}/\Lambda \to \mathbf{P}_1 = \mathbf{C} \cup \{\infty\}$$

defined by $\tilde{\wp}(\Lambda + z) = \wp(z)$ is holomorphic in the sense of Riemann surfaces.

Given that $\wp'(z)$ has a simple zero at each point in $\frac{1}{2}\Lambda - \Lambda$ and no other zeros in $\mathbf{C}$, show that if Λ is generated by ω_1 and ω_2 then

$$(\wp(z) - \wp(\tfrac{1}{2}\omega_1))(\wp(z) - \wp(\tfrac{1}{2}\omega_2))(\wp(z) - \wp(\tfrac{1}{2}\omega_1 + \tfrac{1}{2}\omega_2))/\wp'(z)^2$$

is a holomorphic doubly periodic function on $\mathbf{C}$. Deduce using exercise 5.9 that there is a cubic polynomial $Q(x)$ such that

$$\wp'(z)^2 = Q(\wp(z))$$

for all z.

5.11. Let f be a nonconstant meromorphic function on $\mathbf{C}$ which is even (i.e. $f(-z) = f(z)$ for all $z \in \mathbf{C}$) and doubly periodic with period lattice Λ (i.e.

$f(z + \lambda) = f(z)$ for all $z \in \mathbf{C}$ and $\lambda \in \Lambda$). Let $\wp(z)$ be the Weierstrass $\wp$-function associated to Λ. Using Cauchy's residue theorem applied to the boundary of a suitable parallelogram (as in the proof of proposition 5.18) or otherwise, show that the number $n(f)$ of cosets $\Lambda + a$ consisting of zeros of $f(z)$ is even and positive and is the same as the number of cosets consisting of poles (allowing for multiplicities of zeros and poles). Show that if a is a zero and b is a pole then

$$g(z) = \begin{cases} f(z)(\wp(z) - \wp(b))/(\wp(z) - \wp(a)) & a \notin \Lambda, \ b \notin \Lambda \\ f(z)(\wp(z) - \wp(b)) & a \in \Lambda, \ b \notin \Lambda \\ f(z)/\wp(z) - \wp(a) & a \notin \Lambda, \ b \in \Lambda \end{cases}$$

is an even doubly periodic meromorphic function and either $g(z)$ is constant or $n(g) = n(f) - 2$. Deduce that there is a rational function $R(z)$ such that

$$f(z) = R(\wp(z)).$$

Show that $\wp'(z)^2$ is an even doubly periodic meromorphic function and deduce that

$$\wp'(z)^2 = 4(\wp(z) - \wp(a))(\wp(z) - \wp(b))(\wp(z) - \wp(c))$$

for some $a, b, c \in \mathbf{C} - \Lambda$.

5.12. Let Λ be a lattice in $\mathbf{C}$. Show that

$$g_2(\Lambda)^3 - 27g_3(\Lambda)^2 \neq 0.$$

5.13. Let Λ and $\tilde{\Lambda}$ be lattices in $\mathbf{C}$ and let

$$F : \mathbf{C}/\Lambda \to \mathbf{C}/\tilde{\Lambda}$$

be a holomorphic map. Show that there exist $a, b \in \mathbf{C}$ such that $a\Lambda \subseteq \tilde{\Lambda}$ and

$$F(\Lambda + z) = \tilde{\Lambda} + az + b$$

for all $z \in \mathbf{C}$. [Hint: define maps π and $\tilde{\pi}$ from $\mathbf{C}$ to $\mathbf{C}/\Lambda$ and $\mathbf{C}/\tilde{\Lambda}$ by $\pi(z) = \Lambda + z$, $\tilde{\pi}(z) = \tilde{\Lambda} + z$. If ψ is a holomorphic local inverse for $\tilde{\pi}$ show that the composition $\tilde{\psi} \circ F \circ \pi$ is holomorphic and depends on the choice of $\tilde{\psi}$ only up to the addition of a constant. Deduce that its derivative extends to a doubly periodic holomorphic function on $\mathbf{C}$, and hence is constant].

5.14. Let C and $\tilde{C}$ be nonsingular projective cubic curves in $\mathbf{P}_2$ defined by the equations

$$y^2 z = 4x^3 - g_2 x z^2 - g_3 z^3, \quad y^2 z = 4x^3 - \tilde{g}_2 x z^2 - \tilde{g}_3 z^3.$$

Show that there is a projective transformation of $\mathbf{P}_2$ given by a diagonal matrix taking C to $\tilde{C}$ if and only if $J(C) = J(\tilde{C})$ where

$$J(C) = \frac{g_2^3}{g_2^3 - 27g_3^2}.$$

5.15. Let Λ and $\tilde{\Lambda}$ be lattices in $\mathbf{C}$. Use exercises 5.13 and 5.14 to show that the following statements are equivalent.

(i). $\mathbf{C}/\Lambda$ is biholomorphic to $\mathbf{C}/\tilde{\Lambda}$.
(ii). $\Lambda = a\tilde{\Lambda}$ for some $a \in \mathbf{C} - \{0\}$.
(iii). $J(\Lambda) = J(\tilde{\Lambda})$ where $J(\Lambda) = g_2(\Lambda)^3/(g_2(\Lambda)^3 - 27g_3(\Lambda)^2)$.

[Hint: show (i)$\Rightarrow$(ii)$\Rightarrow$(iii)$\Rightarrow$(i)].

5.16. Let $f : R \to S$ be a holomorphic map between Riemann surfaces with $p \in R$ and $f(p) = q \in S$. Show that there are holomorphic charts $\phi_\alpha : U_\alpha \to V_\alpha$ on R near p and $\psi_\beta : W_\beta \to X_\beta$ on S near q such that

$$\phi_\alpha(p) = 0 = \psi_\beta(q)$$

and $\psi_\beta \circ f \circ \phi_\alpha^{-1}$ is either identically zero or is given by

$$z \mapsto z^\nu$$

for some positive integer ν. Show also that ν is independent of the choice of charts ϕ_α and ψ_β; it is called the *ramification index* of f at p. Show that this definition agrees with the definition given in §4.2. We also say that f takes the value q at p with *multiplicity ν*.

5.17. Let $f : R \to S$ be a nonconstant holomorphic map between connected compact Riemann surfaces. Show that f takes each value in S the same number of times, counting multiplicities. [Hint: use e.g. [Priestley 85] 10.22 to show that if $q \in S$ and $f^{-1}\{q\} = \{p_1, \ldots, p_l\}$ where the multiplicity of f at p_i is ν_i then there are disjoint open neighbourhoods $U_1, \ldots, U_l$ of $p_1, \ldots, p_l$ in R and V of q in S such that the restriction of f to U_i takes each value in V exactly ν_i times, counting multiplicities. Since

$$R - (U_1 \cup \ldots \cup U_n)$$

is compact, deduce that

$$W = V \cap (S - f(R - (U_1 \cup \ldots \cup U_n)))$$

is an open neighbourhood of q in S such that f takes each value in W exactly

$$\nu_1 + \cdots + \nu_l$$

times. Now use the connectedness of S to obtain the required result].

5.18. Let

$$C = \{[x, y, z] \in \mathbf{P}_2 \ : \ P(x, y, z) = 0\}$$

be a nonsingular projective curve. The mapping

$$[x, y, z] \mapsto [\frac{\partial P}{\partial x}, \frac{\partial P}{\partial y}, \frac{\partial P}{\partial z}]$$

from $\mathbf{P}_2$ to $\mathbf{P}_2$ is called the *polar mapping* associated with C, and the image of C is called the *dual curve* $\hat{C}$ of C.

(i). Use Euler's relation (lemma 2.32) and the fact that C is nonsingular to show that the polar mapping is well-defined.

(ii). Show that if C is a conic then the polar mapping is a projective transformation and hence the dual curve $\hat{C}$ is also a nonsingular conic.

(iii). Show that the polar mapping from C to $\tilde{C}$ is holomorphic, but that it does not have a holomorphic inverse if the degree of C is at least three. [Hint: you may wish to consider the inflection points of C].

Chapter 6

Differentials on Riemann surfaces

6.1 Holomorphic differentials

We have associated to a lattice Λ in $\mathbf{C}$ a nonsingular cubic curve C_Λ in $\mathbf{P}_2$ defined by

$$y^2 z = 4x^3 - g_2(\Lambda)xz^2 - g_3(\Lambda)z^3.$$

Conversely given the curve C_Λ can we recover the lattice Λ? For this we need the concept of the integral of a holomorphic differential along a piecewise-smooth path in a Riemann surface.

The definition of a piecewise-smooth path in a Riemann surface is the obvious generalisation of the definition of a piecewise-smooth path in $\mathbf{C}$.

Definition 6.1 *A* piecewise-smooth path *in a Riemann surface S is a continuous map γ from a closed interval $[a, b]$ in $\mathbf{R}$ to S such that if $\phi : U \to V$ is a holomorphic chart on an open subset U of S and $[c, d] \subseteq \gamma^{-1}(U)$ then*

$$\phi \circ \gamma : [c, d] \to V$$

is a piecewise-smooth path in the open subset V of $\mathbf{C}$. (See the definition after remark 5.1). The path γ is closed *if $\gamma(a) = \gamma(b)$.*

The definitions of holomorphic and meromorphic differentials on a Riemann surface are not quite so immediate and need some motivation. First we need to say what we mean by meromorphic functions on a Riemann surface.

Recall from example 5.40(d) that a meromorphic function on an open subset W of $\mathbf{C}$ can be interpreted as a function $W \to \mathbf{P}_1$ which is holomorphic in the sense of Riemann surfaces (and is not identically ∞ on any connected component of W). Here as usual we identify $\mathbf{P}_1$ with $\mathbf{C} \cup \{\infty\}$. Thus it is reasonable to make the following definition.

Definition 6.2 *A meromorphic function on a Riemann surface S is a function $f : S \to \mathbf{P}_1$ which is holomorphic in the sense of Riemann surfaces and is not identically ∞ on any connected component of S.*

Remark 6.3 If we are considering *compact* Riemann surfaces S (such as nonsingular projective curves) then meromorphic functions on S are much more interesting than holomorphic functions $f : S \to \mathbf{C}$. This is because every holomorphic function $f : S \to \mathbf{C}$ on a compact Riemann surface is constant (see exercise 5.4 and lemma 6.26) but there are lots of meromorphic functions $f : S \to \mathbf{P}_1$. This is very hard to prove in general (see e.g. [Jones 71], Chapter VI or [Springer 57]) but we can give examples such as the Weierstrass $\wp$-function on a complex torus, and rational functions on nonsingular projective curves. By a *rational function* on $C - Sing(C)$ where

$$C = \{[x, y, z] \in \mathbf{P}_2 \ : \ P(x, y, z) = 0\}$$

is an irreducible projective curve in $\mathbf{P}_2$, we mean a meromorphic function on $C - Sing(C)$ of the form

$$[x, y, z] \mapsto \frac{S(x, y, z)}{T(x, y, z)}$$

where $S(x, y, z)$ and $T(x, y, z)$ are homogeneous polynomials of the same degree and $T(x, y, z)$ does not vanish identically on C (i.e. $T(x, y, z)$ is not divisible by $P(x, y, z)$). Note that if $S(x, y, z)$ and $T(x, y, z)$ have degree k then

$$\frac{S(\lambda x, \lambda y, \lambda z)}{T(\lambda x, \lambda y, \lambda z)} = \frac{\lambda^k S(x, y, z)}{\lambda^k T(x, y, z)}$$

so the function is independent of the choice of homogeneous coordinates.

Definition 6.4 *Let f and g be meromorphic functions on a Riemann surface S. Then we say that*

$$f \, dg$$

is a meromorphic differential on S, and if $\tilde{f}$ and $\tilde{g}$ are also meromorphic functions on S, we say that

$$f \, dg = \tilde{f} \, d\tilde{g}$$

if and only if for every holomorphic chart $\phi : U \to V$ on an open subset U of S we have

$$(f \circ \phi^{-1})(g \circ \phi^{-1})' = (\tilde{f} \circ \phi^{-1})(\tilde{g} \circ \phi^{-1})'.$$

Note that the compositions of $f, g, \tilde{f}$ and $\tilde{g}$ with ϕ^{-1} are meromorphic functions on the open subset V of $\mathbf{C}$, so we can differentiate them and multiply them together pointwise in the usual way.

Remarks 6.5 (i) Formally a meromorphic differential on S is an equivalence class of pairs (f, g) of meromorphic functions on S such that two pairs (f, g) and $(\tilde{f}, \tilde{g})$ are equivalent if and only if for every holomorphic chart $\phi : U \to V$ on S and for every $z \in V$ we have

$$f \circ \phi^{-1}(z)(g \circ \phi^{-1})'(z) = \tilde{f} \circ \phi^{-1}(z)(\tilde{g} \circ \phi^{-1})'(z).$$

(ii) To get some idea of why this definition might be useful, consider the following question.

"Does it make sense to differentiate a meromorphic function on a Riemann surface S?"

Note that if

$$\Phi = \{\phi_\alpha : U_\alpha \to V_\alpha \ : \ \alpha \in A\}$$

is a holomorphic atlas on S then a meromorphic function $g : S \to \mathbf{C} \cup \{\infty\}$ is determined by the collection of meromorphic functions

$$\{g \circ \phi_\alpha^{-1} : V_\alpha \to \mathbf{C} \cup \{\infty\} \ : \ \alpha \in A\}$$

and conversely a collection of meromorphic functions

$$\{G_\alpha : V_\alpha \to \mathbf{C} \cup \{\infty\} \ : \ \alpha \in A\}$$

defines a meromorphic function g on S such that $G_\alpha = g \circ \phi_\alpha^{-1}$ for each $\alpha \in A$ *if and only if*

$$G_\alpha(\phi_\alpha(u)) = G_\beta(\phi_\beta(u)) \quad \forall u \in U_\alpha \cap U_\beta$$

whenever $\alpha, \beta \in A$.

We can of course differentiate the meromorphic functions $g \circ \phi_\alpha^{-1} : V_\alpha \to \mathbf{C} \cup \{\infty\}$ to get meromorphic functions $(g \circ \phi_\alpha^{-1})' : V_\alpha \to \mathbf{C} \cup \{\infty\}$. However in general these do not define a meromorphic function on S since the compositions $(g \circ \phi_\alpha^{-1})' \circ \phi_\alpha$ and $(g \circ \phi_\beta^{-1})' \circ \phi_\beta$ do not necessarily agree on $U_\alpha \cap U_\beta$. Instead the chain rule tells us that if $u \in U_\alpha \cap U_\beta$ then

$$\begin{aligned} (g \circ \phi_\alpha^{-1})'(\phi_\alpha(u)) &= ((g \circ \phi_\beta^{-1}) \circ (\phi_\beta \circ \phi_\alpha^{-1}))'(\phi_\alpha(u)) \\ &= (g \circ \phi_\beta^{-1})'(\phi_\beta(u))(\phi_\beta \circ \phi_\alpha^{-1})'(\phi_\alpha(u)). \end{aligned}$$

Therefore $(g \circ \phi_\alpha^{-1})' \circ \phi_\alpha$ and $(g \circ \phi_\beta^{-1})' \circ \phi_\beta$ differ by a factor of $(\phi_\beta \circ \phi_\alpha^{-1})' \circ \phi_\alpha$ where $\phi_\beta \circ \phi_\alpha^{-1}$ are the transition functions of the atlas.

This means that if we "differentiate" the meromorphic function g on S by differentiating all the meromorphic functions $g \circ \phi_\alpha^{-1} : V_\alpha \to \mathbf{C}$ which are defined on open subsets of $\mathbf{C}$ and then transferring back to get meromorphic functions $(g \circ \phi_\alpha^{-1})' \circ \phi_\alpha$ on the open subsets U_α which cover S, we do *not* get a *function* on S. Instead, we get an abstract object called a (meromorphic) differential, written dg. We multiply differentials by meromorphic functions f on S to get new differentials written $f\,dg$. Two meromorphic differentials $f\,dg$ and $\tilde{f}\,d\tilde{g}$ on S will then be the same if the meromorphic functions $(f \circ \phi_\alpha^{-1})(g \circ \phi_\alpha^{-1})'$ and $(\tilde{f} \circ \phi_\alpha^{-1})(\tilde{g} \circ \phi_\alpha^{-1})'$ which represent them on U_α via the chart $\phi_\alpha \to V_\alpha$ are equal for all $\alpha \in A$.

These ideas lead to the following alternative definition of a meromorphic differential.

Definition 6.6 *Let $\{\phi_\alpha : U_\alpha \to V_\alpha : \alpha \in A\}$ be a holomorphic atlas on a Riemann surface S. Then a meromorphic differential η on S is given by a collection*

$$\{\eta_\alpha : V_\alpha \to \mathbf{C} \cup \{\infty\} : \alpha \in A\}$$

of meromorphic functions on the open subsets V_α of $\mathbf{C}$ such that if $\alpha, \beta \in A$ and $u \in U_\alpha \cap U_\beta$ then

$$\eta_\alpha(\phi_\alpha(u)) = \eta_\beta(\phi_\beta(u))(\phi_\beta \circ \phi_\alpha^{-1})'(\phi_\alpha(u)).$$

Given two meromorphic functions f and g on S we can define a meromorphic differential $f\,dg$ on S in this sense by $f\,dg = \eta$ where

$$\eta_\alpha = (f \circ \phi_\alpha^{-1})(g \circ \phi_\alpha^{-1})'.$$

Remark 6.7 If η and ζ are meromorphic differentials according to this definition and ζ is not identically zero on any connected component of S then the ratios η_α/ζ_α define meromorphic functions on the open subsets V_α of $\mathbf{C}$ satisfying

$$\eta_\alpha(\phi_\alpha(u))/\zeta_\alpha(\phi_\alpha(u)) = \eta_\beta(\phi_\beta(u))/\zeta_\beta(\phi_\beta(u))$$

for all $u \in U_\alpha$; or equivalently $\eta = f\zeta$. Therefore to show that every meromorphic differential η in the sense of definition 6.6 is a meromorphic differential of the form $f\,dg$ it is enough to show that there is at least one nonconstant meromorphic function g on every Riemann surface. We have already remarked at 6.3 that it is beyond the scope of this book to prove that this is true for any Riemann surface but we have already seen examples of nonconstant meromorphic functions on the Riemann surfaces we are interested in, namely nonsingular curves and complex tori.

Remark 6.8 Note that when S is the complex plane $\mathbf{C}$ any meromorphic differential $f\,dg$ on S can be expressed uniquely in the form $h\,dz$ where $h = fg'$ is meromorphic and z denotes the identity function on $\mathbf{C}$. Thus meromorphic differentials on $\mathbf{C}$ can be identified naturally with meromorphic functions on $\mathbf{C}$.

Remark 6.9 The most important thing to remember about a meromorphic differential $f\,dg$ is that it is represented on the range of a holomorphic chart $\phi_\alpha : U_\alpha \to V_\alpha$ by the meromorphic function $(f \circ \phi_\alpha^{-1})(g \circ \phi_\alpha^{-1})'$.

Definition 6.10 *We say that the meromorphic differential $f\,dg$ has a* pole *at a point p in S if the meromorphic function $(f \circ \phi^{-1})(g \circ \phi^{-1})'$ has a pole at $\phi(p)$ when $\phi : U \to V$ is a holomorphic chart on an open neighbourhood U of p in S. We call $f\,dg$ a* holomorphic differential *if it has no poles[†].*

Our reason for introducing holomorphic differentials is that they are the natural objects to integrate on a Riemann surface S.

Definition 6.11 *If $f\,dg$ is a holomorphic differential on S then the* integral *of $f\,dg$ along a piecewise-smooth path $\gamma : [a, b] \to S$ is*

$$\int_\gamma f\,dg = \int_a^b f \circ \gamma(t)(g \circ \gamma)'(t)dt.$$

Remark 6.12 To know that this definition makes sense we must check that if $f\,dg = \tilde{f}\,d\tilde{g}$ then

$$\int_a^b f \circ \gamma(t)(g \circ \gamma)'(t)dt = \int_a^b \tilde{f} \circ \gamma(t)(\tilde{g} \circ \gamma)'(t)dt.$$

We know that

$$(f \circ \phi_\alpha^{-1})(g \circ \phi_\alpha^{-1})' = (\tilde{f} \circ \phi_\alpha^{-1})(\tilde{g} \circ \phi_\alpha^{-1})'$$

for every holomorphic chart $\phi_\alpha : U_\alpha \to V_\alpha$. Since the open subsets U_α cover S we can find

$$a = a_0 < a_1 < \ldots < a_p = b$$

and $\alpha_1, \ldots, \alpha_p \in A$ such that

$$\gamma([a_{i-1}, a_i]) \subseteq U_{\alpha_i}$$

if $1 \leq i \leq p$. Then by the chain rule

$$\int_\gamma f\,dg = \sum_{i=1}^p \int_{a_{i-1}}^{a_i} f \circ \gamma(t)(g \circ \gamma)'(t)dt$$

$$= \sum_{i=1}^p \int_{a_{i-1}}^{a_i} (f \circ \phi_{\alpha_i}^{-1}) \circ (\phi_{\alpha_i} \circ \gamma)(t)(g \circ \phi_{\alpha_i}^{-1})' \circ (\phi_{\alpha_i} \circ \gamma)(t)(\phi_{\alpha_i} \circ \gamma)'(t)dt$$

and this last expression is unaltered if we replace f by $\tilde{f}$ and g by $\tilde{g}$.

[†] This is well defined: see the discussion on page 160 below.

Remark 6.13 If $\psi : [c, d] \to [a, b]$ is a piecewise-smooth map between the intervals $[c, d]$ and $[a, b]$ then $\gamma \circ \psi : [c, d] \to S$ is a piecewise-smooth path in S and on substituting $t = \psi(s)$ we find

$$\int_\gamma f \, dg = \int_c^d f \circ \gamma \circ \psi(s)(g \circ \gamma)'(\psi(s))\psi'(s)ds = \int_{\gamma \circ \psi} f \, dg.$$

This tells us that the integral of a differential along a path is independent of the parametrisation of the path.

Examples 6.14 (1). If the Riemann surface S is $\mathbf{C}$ then

$$\int_\gamma f \, dg = \int_\gamma f(z)g'(z)dz$$

is the integral of $f(z)g'(z)$ along γ in the usual sense of complex analysis.

(2). If $g : S \to \mathbf{C}$ is a complex-valued holomorphic mapping on any Riemann surface S then

$$\int_\gamma dg = g(\gamma(b)) - g(\gamma(a)).$$

Definition 6.15 *If $\psi : S \to R$ is a holomorphic mapping between Riemann surfaces S and R, and if $f \, dg$ is a holomorphic differential on R then we define a holomorphic differential $\psi^*(f \, dg)$ on S by*

$$\psi^*(f \, dg) = (f \circ \psi) \, d(g \circ \psi).$$

Then if $\gamma : [a, b] \to S$ is a piecewise-smooth path in S we have

$$\int_\gamma \psi^*(f \, dg) = \int_a^b f \circ \psi \circ \gamma(t)(g \circ \psi \circ \gamma)'(t)dt = \int_{\psi \circ \gamma} f \, dg.$$

Digression on elliptic, hyperelliptic and abelian integrals 6.16
Let

$$C = \{[x, y, z] \in \mathbf{P}_2 \; : \; P(x, y, z) = 0\}$$

be an irreducible projective curve in $\mathbf{P}_2$. An *abelian integral* is an integral of the form

$$\int_\gamma f \, dg$$

where f and g are rational functions on $C - Sing(C)$ (see remark 6.3) and γ is a piecewise-smooth path in $C - Sing(C)$ not passing through any poles of the meromorphic differential $f \, dg$.

We usually assume that C is not the line at infinity defined by $z = 0$ and take the rational function g to be

$$[x, y, z] \mapsto \frac{x}{z}.$$

We often then work in affine coordinates $[x, y, 1]$ and write dx for dg. In affine coordinates f becomes a rational function $R(x, y)$ of x and y in the usual sense, and our integral is written

$$\int_\gamma f \, dg = \int_\gamma R(x, y) dx$$

where y is regarded as a multivalued function of x via the equation $P(x, y, 1)$ which defines C in affine coordinates.

The integral is called an *elliptic integral* if C is an elliptic curve, i.e. it is defined in appropriate affine coordinates by

$$y^2 = (x - \alpha_1)(x - \alpha_2) \ldots (x - \alpha_k)$$

where $\alpha_1, \ldots, \alpha_k$ are distinct complex numbers and $k = 3$ or 4. If $k \geq 5$ then C is called a hyperelliptic curve and the integral $\int_\gamma f \, dg$ is called a *hyperelliptic integral*.

If the curve γ avoids the points of $\mathbf{C}$ where $y = 0$ or $z = 0$ then the integral becomes

$$\int_{\tilde{\gamma}} R(w, \sqrt{(w - \alpha_1)(w - \alpha_2) \ldots (w - \alpha_k)}) dw$$

in the usual sense of complex analysis, where $\tilde{\gamma}$ is the composition of γ with the local chart $[x, y, 1] \mapsto x$ and an appropriate choice of the square root is made at $\tilde{\gamma}(t)$ for each $t \in [a, b]$. As was mentioned in §1.2.4 integrals of this form were greatly studied in the nineteenth century. It was found that when k is one or two they can be expressed in terms of rational functions and logarithms (see §1.2.4). However when $k \geq 3$ this is no longer the case and the integrals were used to define new functions (called elliptic and hyperelliptic functions). The difference between the two cases arises essentially because an irreducible projective curve of degree one or two in $\mathbf{P}_2$ is isomorphic to the complex projective line $\mathbf{P}_1$ (see corollary 3.12) whereas elliptic and hyperelliptic curves have genus at least one and hence cannot be homeomorphic, let alone isomorphic, to $\mathbf{P}_1$.

Now we return to the study of complex tori and nonsingular cubic curves. Given a lattice Λ in $\mathbf{C}$ we have defined a biholomorphism

$$u : \mathbf{C}/\Lambda \to C_\Lambda$$

(see definition 5.22) where C_Λ is the nonsingular cubic curve in $\mathbf{P}_2$ defined by

$$y^2 z = 4x^3 - g_2(\Lambda)xz^2 - g_3(\Lambda)z^3.$$

As in digression 6.16 there is a meromorphic differential on C_Λ given in inhomogeneous coordinates $[x, y, 1]$ by $y^{-1}dx$. Let

$$\eta = u^*(y^{-1}dx)$$

(see definition 6.15). Then η is a meromorphic differential on $\mathbf{C}/\Lambda$. Moreover if

$$\pi : \mathbf{C} \to \mathbf{C}/\Lambda$$

is defined by $\pi(z) = \Lambda + z$ then

$$
\begin{aligned}
\pi^*\eta &= \pi^* u^*(y^{-1}dx) \\
&= (u \circ \pi)^*(y^{-1}dx) \\
&= (\wp')^{-1}d\wp \\
&= (\wp')^{-1}\wp'dz \\
&= dz
\end{aligned}
$$

where $z : \mathbf{C} \to \mathbf{C}$ denotes the identity function. Since π is locally a holomorphic bijection with holomorphic inverse and dz is a holomorphic differential on $\mathbf{C}$ it follows that η has no poles, i.e. η is a holomorphic differential on $\mathbf{C}/\Lambda$. Since u is a holomorphic bijection with holomorphic inverse it also follows that $y^{-1}dx$ is a holomorphic differential on C_Λ.

Choose any $\lambda \in \Lambda$ and define $\tilde{\gamma} : [0,1] \to \mathbf{C}$ by $\tilde{\gamma}(t) = t\lambda$. If $\gamma = \pi \circ \tilde{\gamma} : [0,1] \to \mathbf{C}/\Lambda$ then

$$\gamma(t) = \Lambda + t\lambda$$

and hence

$$\gamma(0) = \Lambda + 0 = \Lambda + \lambda = \gamma(1),$$

so γ is a piecewise-smooth closed path in $\mathbf{C}/\Lambda$. By definition 6.15 and example 6.14(2) we have

$$\int_\gamma \eta = \int_{\tilde{\gamma}} \pi^*\eta = \int_{\tilde{\gamma}} dz = \tilde{\gamma}(1) - \tilde{\gamma}(0) = \lambda.$$

On the other hand if $\gamma : [a, b] \to \mathbf{C}/\Lambda$ is any piecewise-smooth closed path in $\mathbf{C}/\Lambda$ then by lemma C.3 of Appendix C we can find a continuous path $\tilde{\gamma} : [a, b] \to \mathbf{C}$ such that $\pi \circ \tilde{\gamma} = \gamma$. This path $\tilde{\gamma}$ is piecewise-smooth because π is locally a holomorphic bijection with a holomorphic inverse. Since

$$\pi \circ \tilde{\gamma}(b) = \gamma(b) = \gamma(a) = \pi \circ \tilde{\gamma}(a)$$

we have

$$\int_\gamma \eta = \int_{\tilde{\gamma}} \pi^* \eta = \int_{\tilde{\gamma}} dz = \tilde{\gamma}(b) - \tilde{\gamma}(a) \in \Lambda.$$

Thus we have proved

Proposition 6.17

$$\Lambda = \{ \int_\gamma \eta \; : \; \gamma \text{ is a closed piecewise} - \text{smooth path in } \mathbf{C}/\Lambda \}.$$

□

Since $u : \mathbf{C}/\Lambda \to C_\Lambda$ is a bijection with a holomorphic inverse and

$$\eta = u^*(y^{-1} dx)$$

we can use definition 6.15 to deduce immediately

Corollary 6.18

$$\Lambda = \{ \int_\gamma y^{-1} dx \; : \; \gamma \text{ is a closed piecewise} - \text{smooth path in } C_\Lambda \}.$$

□

This means that we can recover the lattice Λ from the curve C_Λ in $\mathbf{P}_2$. We can also describe $u^{-1} : C_\Lambda \to \mathbf{C}/\Lambda$ in terms of integrals of the differential $y^{-1} dx$ on C_Λ.

Proposition 6.19 *The inverse of the holomorphic bijection*

$$u : \mathbf{C}/\Lambda \to C_\Lambda$$

is given by

$$u^{-1}(p) = \Lambda + \int_{[0,1,0]}^{p} y^{-1} dx$$

where the integral is over any piecewise-smooth path γ in C_Λ from $[0,1,0]$ to p.

Remark 6.20 This makes sense because if γ_1 and γ_2 are both piecewise-smooth paths in C_Λ from $[0,1,0]$ to p then

$$\int_{\gamma_1} y^{-1} dx - \int_{\gamma_2} y^{-1} dx$$

is the integral of $y^{-1} dx$ along a *closed* piecewise-smooth path in C_Λ and hence belongs to Λ by corollary 6.18.

Proof of proposition 6.19. Suppose that $u(\Lambda+z) = p$ and that $\gamma : [a, b] \to C_\Lambda$ is any piecewise-smooth path from $[0, 1, 0]$ to p. Then $u^{-1} \circ \gamma$ is a path in $\mathbf{C}/\Lambda$ from $\Lambda + 0$ to $\Lambda + z$ and hence

$$
\begin{aligned}
\int_\gamma y^{-1}dx &= \int_{u^{-1}\circ\gamma} u^*(y^{-1}dx) \\
&= \int_{u^{-1}\circ\gamma} \eta \\
&= \int_{\tilde\gamma} \pi^*\eta \\
&= \tilde\gamma(b) - \tilde\gamma(a)
\end{aligned}
$$

where $\tilde\gamma : [a, b] \to \mathbf{C}$ is any piecewise-smooth path in $\mathbf{C}$ such that $\pi \circ \tilde\gamma = u^{-1}\circ\gamma$. Then $\Lambda+\tilde\gamma(b) = u^{-1}(p) = \Lambda+z$ and also $\Lambda+\tilde\gamma(a) = u^{-1}[0, 1, 0] = \Lambda+0$ so

$$
\Lambda + \int_\gamma y^{-1}dx = \Lambda + \tilde\gamma(b) - \tilde\gamma(a) = \Lambda + z
$$

as required.

6.2   Abel's theorem

In §5.2 we saw that any complex torus $\mathbf{C}/\Lambda$ is biholomorphic to a nonsingular cubic curve C_Λ in $\mathbf{P}_2$. Since $\mathbf{C}/\Lambda$ is an abelian group under addition the cubic curve C_Λ has an induced abelian group structure. We can describe this group structure in terms of the geometry of the curve C_Λ in $\mathbf{P}_2$. It is determined completely by two properties: firstly that the zero (identity) element is the inflection point $[0, 1, 0]$ and secondly that three points p, q, r add up to zero if and only if they are the three points of intersection (allowing for multiplicities) of the cubic C_Λ with a line in $\mathbf{P}_2$. This last fact (which gives an addition formula for elliptic integrals) is called *Abel's theorem* for the cubic (theorem 6.23 below) and the aim of this section is to give a proof of it.

Remark 6.21 We have already observed (theorem 3.38) that if p_0 is a point of inflection on a nonsingular projective cubic curve C then there is a unique additive group structure on C such that the zero element is p_0 and three points add up to zero if and only if they are the points of intersection (allowing for multiplicities) of C with a line in $\mathbf{P}_2$. Given this, Abel's theorem can be regarded as the statement that the holomorphic bijection

$$
u : \mathbf{C}/\Lambda \to C_\Lambda
$$

defined at 5.22 is a group isomorphism with respect to this group structure (with $p_0 = [0, 1, 0]$) on C_Λ and the quotient group structure on $\mathbf{C}/\Lambda$. However we did not give a complete proof that an additive group structure with the required properties exists on C; the law of associativity was left unproved. Abel's theorem gives an alternative indirect proof of the existence of an additive group structure with required properties on cubic curves of the form C_Λ; we can simply use the bijection u to transfer the additive group structure on $\mathbf{C}/\Lambda$ to one on C_Λ and Abel's theorem tells us it has the properties we want. In fact (see exercises 6.13 and 6.14) every nonsingular projective cubic curve in $\mathbf{P}_2$ is equivalent under a projective transformation to one of the form C_Λ for some lattice Λ in $\mathbf{C}$.

Remark 6.22 Recall from lemma 3.37 that a line in $\mathbf{P}_2$ meets a nonsingular projective cubic curve C in $\mathbf{P}_2$ *either*

(a) in three distinct points p, q, r each with multiplicity one (i.e. L is not the tangent line to C at p, q or r), *or*

(b) in two distinct points p with multiplicity one and q with multiplicity two (i.e. L is the tangent line to C at q but not at p, and q is not an inflection point of C), *or*

(c) in one point p with multiplicity three (i.e. L is the tangent line to C at p and p is an inflection point of C).

This means that the group structure on C_Λ is such that if p, q, r are *distinct* points on C_Λ then

$$p + q + r = 0$$

if and only if p, q, r all lie on a line in $\mathbf{P}_2$,

$$p + p + q = 0$$

if and only if the tangent to C_Λ at p passes through q, and

$$p + p + p = 0$$

if and only if p is a point of inflection on C_Λ. In particular we see that the points of inflection on C_Λ are precisely the points of order 1 or 3. Under the group isomorphism $u : \mathbf{C}/\Lambda \to C_\Lambda$ these correspond to the points of order 1 or 3 in $\mathbf{C}/\Lambda$. There are precisely nine such points in $\mathbf{C}/\Lambda$; they can be written in the form

$$\Lambda + \frac{j}{3}\omega_1 + \frac{k}{3}\omega_2$$

where $j, k \in \{0, 1, 2\}$. Thus these points form a subgroup of $\mathbf{C}/\Lambda$ isomorphic to the product of two copies of the cyclic group of order three. Moreover they can be arranged in a 3×3 array (see figure 6.1) such that the three entries in any row, column or diagonal add up to $\Lambda + 0$. It follows that C_Λ has exactly nine points of inflection p_{jk} for $j, k \in \{0, 1, 2\}$ (cf. example 3.36), which can

$$\Lambda + 0 \qquad\qquad \Lambda + \tfrac{1}{3}\omega_2 \qquad\qquad \Lambda + \tfrac{2}{3}\omega_2$$

$$\Lambda + \tfrac{1}{3}\omega_1 \qquad \Lambda + \tfrac{1}{3}\omega_1 + \tfrac{1}{3}\omega_2 \qquad \Lambda + \tfrac{1}{3}\omega_1 + \tfrac{2}{3}\omega_2$$

$$\Lambda + \tfrac{2}{3}\omega_1 \qquad \Lambda + \tfrac{2}{3}\omega_1 + \tfrac{1}{3}\omega_2 \qquad \Lambda + \tfrac{2}{3}\omega_1 + \tfrac{2}{3}\omega_2$$

Figure 6.1: The points of order 1 or 3 in $\mathbf{C}/\Lambda$

$$p_{00} \qquad\qquad p_{01} \qquad\qquad p_{02}$$

$$p_{10} \qquad\qquad p_{11} \qquad\qquad p_{12}$$

$$p_{20} \qquad\qquad p_{21} \qquad\qquad p_{22}$$

Figure 6.2: The points of inflection on C_Λ

be arranged in a 3×3 array (see figure 6.2) such that the three points in any row, column or diagonal all lie on a line in $\mathbf{P}_2$. Here the term "diagonal" includes for example p_{01}, p_{12}, p_{20}. Pictorially this can be expressed as in figure 6.3. Moreover the nine points of inflection of C_Λ form a subgroup isomorphic to the product of two copies of the cyclic group of order three.

In order to state Abel's theorem, let us recall from proposition 6.19 that if Λ is a lattice in $\mathbf{C}$ and C_Λ is the curve in $\mathbf{P}_2$ defined by the polynomial

$$Q_\Lambda(x,y,z) = y^2 z - 4x^3 + g_2(\Lambda)xz^2 + g_3(\Lambda)z^3$$

and $u : \mathbf{C}/\Lambda \to C_\Lambda$ is defined as at (5.22) by

$$u(\Lambda + z) = \begin{cases} [\wp(z), \wp'(z), 1] & \text{if } z \notin \Lambda \\ [0, 1, 0] & \text{if } z \in \Lambda, \end{cases}$$

then u is a holomorphic bijection and its inverse is given by

$$u^{-1}(p) = \Lambda + \int_{[0,1,0]}^{p} y^{-1} dx$$

where the integral is over any piecewise-smooth path in C_Λ from $[0,1,0]$ to p.

Theorem 6.23 Abel's theorem for tori.

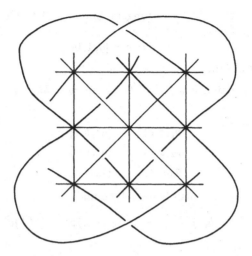

Figure 6.3: The lines joining the points of inflection on C_Λ

If $t, v, w \in \mathbf{C}$ then

$$t + v + w \in \Lambda$$

if and only if there is a line L in $\mathbf{P}_2$ whose intersection with C_Λ consists of the points $u(\Lambda + t), u(\Lambda + v)$ and $u(\Lambda + w)$ (allowing for multiplicities). Equivalently if $p, q, r \in C_\Lambda$ then

$$\Lambda + \int_{[0,1,0]}^{p} y^{-1} dx + \int_{[0,1,0]}^{q} y^{-1} dx + \int_{[0,1,0]}^{r} y^{-1} dx = \Lambda + 0$$

if and only if p, q, r are the points of intersection of C_Λ with a line in $\mathbf{P}_2$.

Remark 6.24 We can interpret Abel's theorem as an addition formula modulo Λ for elliptic integrals of the form

$$\int_{[0,1,0]}^{p} y^{-1} dx$$

on C_Λ.

Proof of Abel's theorem. First let us show that if L is a line in $\mathbf{P}_2$ which intersects C_Λ in the points p, q, r (allowing for multiplicities) then

$$\Lambda + \int_{[0,1,0]}^{p} y^{-1} dx + \int_{[0,1,0]}^{q} y^{-1} dx + \int_{[0,1,0]}^{r} y^{-1} dx = \Lambda + 0.$$

We consider three cases, each more general than the last.

Case 1. Suppose that L is the tangent line $z = 0$ to C_Λ at the point of inflection $[0, 1, 0]$. Then

$$p = q = r = [0, 1, 0]$$

so the equality we want to prove is trivial.

Case 2. Suppose that L is a line of the form $cy = bz$. Then L meets C_Λ in the three points

$$\begin{aligned}
p_1(b, c) &= [a_1, b, c] \\
p_2(b, c) &= [a_2, b, c] \\
p_3(b, c) &= [a_3, b, c]
\end{aligned}$$

where a_1, a_2, a_3 are the roots of the polynomial

$$Q_\Lambda(x, b, c) = b^2 c - 4x^3 + g_2(\Lambda)xc^2 + g_3(\Lambda)c^3$$

in x. Define a map

$$\mu : \mathbf{P}_1 \to \mathbf{C}/\Lambda$$

by

$$\mu[b, c] = \Lambda + \int_{[0,1,0]}^{p_1(b,c)} y^{-1}dx + \int_{[0,1,0]}^{p_2(b,c)} y^{-1}dx + \int_{[0,1,0]}^{p_3(b,c)} y^{-1}dx$$

where the integrals are over any paths in $\mathbf{C}$ from $[0, 1, 0]$ to $p_1(b, c), p_2(b, c)$ and $p_3(b, c)$. This map μ is well-defined by remark 6.20. We need two lemmas.

Lemma 6.25 $\mu : \mathbf{P}_1 \to \mathbf{C}/\Lambda$ *is holomorphic.*

Lemma 6.26 *Any holomorphic map from* $\mathbf{P}_1$ *to* $\mathbf{C}$ *is constant.*

Given these lemmas (whose proofs are at the end of this section) we deduce from lemma C.3 of Appendix C that there is a continuous map $\tilde{\mu} : \mathbf{P}_1 \to \mathbf{C}$ such that $\mu = \pi \circ \tilde{\mu}$, where as before π from $\mathbf{C}$ to $\mathbf{C}/\Lambda$ is defined by $\pi(z) = \Lambda + z$. Since the inverse of the restriction of π to a suitable open neighborhood of any $a \in \mathbf{C}$ defines a holomorphic chart on a neighbourhood of $\Lambda + a$ in $\mathbf{C}/\Lambda$ and μ is holomorphic it follows that $\tilde{\mu}$ is holomorphic, and hence by lemma 6.26 $\tilde{\mu}$ is constant. Thus μ is a constant map. Since $\mu[1, 0] = \Lambda + 0$ by Case 1 it follows that $\mu[b, c] = \Lambda + 0$ for all $[b, c] \in \mathbf{P}_1$, i.e. that the equality we want to prove holds when L is of the form $cy = bz$.

Case 3. Suppose now that L is any line in $\mathbf{P}_2$. Then the equation for L can be written in the form

$$sx + t(cy - bz) = 0$$

for some s, t not both zero and b, c not both zero. Fix b, c and define $\nu : \mathbf{P}_1 \to \mathbf{C}/\Lambda$ by

$$\nu[s, t] = \Lambda + \int_{[0,1,0]}^{q_1(s,t)} y^{-1}dx + \int_{[0,1,0]}^{q_2(s,t)} y^{-1}dx + \int_{[0,1,0]}^{q_3(s,t)} y^{-1}dx$$

where $q_1(s,t), q_2(s,t)$ and $q_3(s,t)$ are the points of intersection of C_Λ with the line

$$sx + t(cy - bz) = 0.$$

As in Case 2 this map ν is holomorphic and hence constant. Moreover we know from Case 2 that $\nu[0,1] = \Lambda + 0$ so that $\nu[s,t] = \Lambda + 0$ for all $[s,t] \in \mathbf{P}_1$. This completes the proof that

$$\Lambda + \int_{[0,1,0]}^p y^{-1}dx + \int_{[0,1,0]}^q y^{-1}dx + \int_{[0,1,0]}^r y^{-1}dx = \Lambda + 0$$

when p, q, r are the points of intersection of C_Λ with a line in $\mathbf{P}_2$ (given lemmas 6.25 and 6.26).

It now follows immediately from the form of the inverse of the map u (see proposition 6.19) that if t, v, w are complex numbers and $u(\Lambda + t), u(\Lambda + v), u(\Lambda + w)$ are the points of intersection of C_Λ with a line in $\mathbf{P}_2$ then

$$t + v + w \in \Lambda.$$

Conversely suppose that $t, v, w \in \mathbf{C}$ and that $t + v + w \in \Lambda$. Let

$$p = u(\Lambda + t), \quad q = u(\Lambda + v), \quad r = u(\Lambda + w).$$

If $p \neq q$ let L be the line in $\mathbf{P}_2$ through p and q, and if $p = q$ let L be the tangent line to C_Λ at p. In both cases L meets C_Λ in p, q and another point $\tilde{r}$. Then by what we have just proved

$$\begin{aligned} u^{-1}(p) + u^{-1}(q) + u^{-1}(\tilde{r}) &= \Lambda + 0 \\ &= \Lambda + t + v + w \\ &= u^{-1}(p) + u^{-1}(q) + u^{-1}(r) \end{aligned}$$

so that $u^{-1}(\tilde{r}) = u^{-1}(r)$ and hence $\tilde{r} = r$.

This completes the proof of Abel's theorem, given lemmas 6.25 and 6.26. $\square$

Proof of lemma 6.25. We have to show that the mapping $\mu : \mathbf{P}_1 \to \mathbf{C}/\Lambda$ defined by

$$\mu[b,c] = \Lambda + \int_{[0,1,0]}^{p_1(b,c)} y^{-1}dx + \int_{[0,1,0]}^{p_2(b,c)} y^{-1}dx + \int_{[0,1,0]}^{p_3(b,c)} y^{-1}dx$$

is holomorphic.

For all but finitely many $b \in \mathbf{C}$ the partial derivative

$$\frac{\partial Q_\Lambda}{\partial x}(x,y,z)$$

of the polynomial Q_Λ defining C_Λ is nonzero at $(a, b, 1)$ when a is a root of the polynomial $Q_\Lambda(x, b, 1)$ in x. For such a, b the polynomial $Q_\Lambda(x, b, 1)$ has three distinct roots a_1, a_2, a_3 say, and

$$p_i(b, 1) = [a_i, b, 1] \quad \text{for } i = 1, 2, 3.$$

By the implicit function theorem (see Appendix B) applied to the polynomial $Q_\Lambda(x, y, 1)$ there are open neighbourhoods U and V_1, V_2, V_3 of b and a_1, a_2, a_3 in C and holomorphic functions $g_i : U \to V_i$ for $i = 1, 2, 3$ such that if $x \in V_i$ and $y \in U$ then

$$Q_\Lambda(x, y, 1) = 0 \quad \longleftrightarrow \quad x = g_i(y).$$

Hence there are holomorphic maps $\psi_i : U \to C_\Lambda$ given by

$$\psi_i(w) = [g_i(w), w, 1].$$

We may choose V_1, V_2, V_3 to be disjoint. This means that if $w \in U$ then $g_1(w), g_2(w)$ and $g_3(w)$ are distinct roots of the polynomial $Q_\Lambda(x, w, 1)$ and so

$$p_i(w, 1) = [g_i(w), w, 1] = \psi_i(w).$$

Thus if γ is a path in U from b to w then $\psi_i \circ \gamma$ is a path in C_Λ from $p_i(b, 1)$ to $p_i(w, 1)$ and

$$\int_{\psi_i \circ \gamma} y^{-1} dx = \int_\gamma \frac{g_i'(w)}{w} dw.$$

Thus

$$\mu[w, 1] = \mu[b, 1] + \sum_{1 \le i \le 3} \int_b^w \frac{g_i'(y)}{y} dy$$

where the integrals are over any path in U from b to w. Since $g_i'(y) = 0$ when $y = 0$ the functions $g_i'(y)/y$ are holomorphic on U so their integrals from b to w are holomorphic functions of w near b. Thus μ is holomorphic in a neighbourhood of $[b, 1]$.

Thus we have shown that μ is holomorphic except possibly at finitely many points of $\mathbf{P}_1$. To show that μ is holomorphic everywhere it suffices by theorem B.6 of Appendix B to show that μ is continuous. This is left as an exercise (see exercise 6.4). $\square$

Proof of lemma 6.26. We need to show that a holomorphic map $f : \mathbf{P}_1 \to C$ is constant. This follows immediately from exercise 5.4 since $\mathbf{P}_1$ is compact. Alternatively we can regard f as a holomorphic map $f : \mathbf{P}_1 \to \mathbf{P}_1$ which never takes the value ∞. Then the result follows from lemma 5.41 which says that if f were nonconstant then it would have to be rational and hence would take the value ∞ somewhere on $\mathbf{P}_1$. $\square$

6.3 The Riemann-Roch theorem

In this section we shall see that the *genus g* of a nonsingular projective curve C in $\mathbf{P}_2$, which we defined in Chapter 4 using the topological properties of C, can also be characterised in terms of the holomorphic structure of C. In fact, g is the dimension of the vector space of holomorphic differentials on C. This is a special case of the famous Riemann-Roch theorem (theorem 6.37 below) which relates dimensions of vector spaces of meromorphic functions with prescribed poles and zeros on C. The Riemann-Roch theorem has many very useful consequences, including an easy proof of the law of associativity for the additive group structure on a nonsingular cubic (see theorem 3.38 and theorem 6.39) and a proof that every meromorphic function on a nonsingular projective curve is rational (theorem 6.41). On the way to proving the Riemann-Roch theorem we shall also find yet another characterisation of the genus g of C, which is that the number of zeros minus the number of poles (counted with multiplicities) of any nonzero meromorphic differential on C is $2g - 2$.

In order to state the Riemann-Roch theorem it is convenient to introduce some new terminology.

Definition 6.27 *A* divisor D *on* C *is a formal sum*

$$D = \sum_{p \in C} n_p\, p$$

such that $n_p \in \mathbf{Z}$ for every $p \in C$ and $n_p = 0$ for all but finitely many $p \in C$. The degree *of D is then*

$$deg(D) = \sum_{p \in C} n_p.$$

If $n_p = 0$ for $p \notin \{p_1, \ldots, p_k\}$ then we also write

$$D = m_1 p_1 + \cdots + m_k p_k$$

where $m_j = n_{p_j}$. By convention we may omit every m_j equal to 1 and write p instead of $1\,p$. If $n_p = 0$ for all $p \in C$ we write

$$D = 0.$$

We add and subtract divisors and multiply them by integers in the obvious way. The set of all divisors on C then becomes an abelian group, denoted $Div(C)$, and the degree defines a homomorphism from $Div(C)$ to $\mathbf{Z}$.

If $n_p \geq 0$ for all $p \in C$ we write $D \geq 0$ and say that D is *effective* (or positive). We write $D \geq D'$ if $D - D' \geq 0$. Note that if $D \geq D'$ then

$$deg\, D \geq deg\, D'.$$

Let f be a meromorphic function on C which is not identically zero. If $p \in C$ we can choose a holomorphic chart $\phi_\alpha : U_\alpha \to V_\alpha$ on C such that $p \in U_\alpha$. Then $f \circ \phi_\alpha^{-1}$ is a meromorphic function on the open subset V_α of $\mathbf{C}$. If m is a positive integer we say that f has a pole or a zero of multiplicity m at p if $f \circ \phi_\alpha^{-1}$ has a pole or a zero of multiplicity m at $\phi_\alpha(p)$, i.e. near $\phi_\alpha(p)$ we can write

$$f \circ \phi_\alpha^{-1}(z) = (z - \phi_\alpha(p))^{-m} g(z)$$

or

$$f \circ \phi_\alpha^{-1}(z) = (z - \phi_\alpha(p))^m g(z)$$

where $g(z)$ is holomorphic and does not vanish at $\phi_\alpha(p)$. This is independent of the choice of chart $\phi_\alpha : U_\alpha \to V_\alpha$, since if $\phi_\beta : U_\beta \to V_\beta$ is another holomorphic chart on C such that $p \in U_\beta$, then the transition function $\phi_\alpha \circ \phi_\beta^{-1}$ satisfies

$$\phi_\alpha \circ \phi_\beta^{-1}(z) = \phi_\alpha(p) + (z - \phi_\beta(p))h(z)$$

where $h(z)$ is holomorphic and does not vanish at $\phi_\beta(p)$.

Similarly if $\omega = f\, dg$ is a meromorphic differential on C which is not identically zero we say that ω has a pole or a zero of multiplicity m at p if the meromorphic function

$$(f \circ \phi_\alpha^{-1})(g \circ \phi_\alpha^{-1})'$$

which represents ω on V_α has a pole or zero of multiplicity m at $\phi_\alpha(p)$. Again this is independent of the choice of chart $\phi_\alpha : U_\alpha \to V_\alpha$.

Definition 6.28 *The divisor of a meromorphic function f which is not identically zero on C is*

$$(f) = \sum_{p \in C} n_p\, p$$

where n_p is $-m$ if f has a pole of multiplicity m at p, is m if f has a zero of multiplicity m at p and is zero otherwise. Note that

$$(fg) = (f) + (g)$$

and

$$\left(\frac{f}{g}\right) = (f) - (g).$$

A divisor which is the divisor of some meromorphic function is called a principal *divisor. Two divisors D and D' are said to be* linearly equivalent, *written*

$$D \sim D',$$

if $D - D'$ is a principal divisor.

If ω is a meromorphic differential on C which is not identically zero then we can define the divisor (ω) of ω using the multiplicities of the poles and zeros of ω in exactly the same way as we defined the divisor of a meromorphic function. The divisor of a meromorphic differential is called a *canonical divisor* and is often written κ. We have already noted in remark 6.7 that if η is another meromorphic differential on C which is not identically zero then there is a meromorphic function f on C such that $\eta = f\omega$, and hence

$$(\eta) = (f) + (\omega) \sim (\omega).$$

Thus any two canonical divisors are linearly equivalent.

At this point we need a result whose proof will be left to the end of the section.

Proposition 6.29 *A principal divisor on C has degree zero; that is, a mero-morphic function on C which is not identically zero has the same number of zeros and poles, counted with multiplicities.*

This has an immediate corollary.

Corollary 6.30 *Two linearly equivalent divisors on C have the same degree. In particular canonical divisors on C all have the same degree.*

Proof. This is a direct consequence of proposition 6.29. □

The proof of the next result will also be postponed to the end of the section.

Proposition 6.31 *If κ is a canonical divisor on C then*

$$deg\ \kappa = 2g - 2.$$

Remark 6.32 Of course this proposition gives us an alternative characterisation of the genus of C; it is one more than half the number of zeros minus the number of poles counted with multiplicities of any meromorphic differential on C which is not identically zero.

Remark 6.33 The proof we shall give of proposition 6.31 depends on the fact that C is a nonsingular projective curve in $\mathbf{P}_2$ and uses the degree-genus formula. However it is possible to prove this result (and indeed the Riemann-Roch theorem) for an arbitrary compact connected Riemann surface (see e.g. [Griffiths & Harris 78])

The vector spaces of meromorphic functions on C which appear in the statement of the Riemann-Roch theorem are defined as follows.

Definition 6.34 *Let $D = \sum_{p \in C} n_p\, p$ be a divisor on C; then $\mathcal{L}(D)$ is the set of meromorphic functions on C satisfying*

$$(f) + D \geq 0$$

together with the zero function. In other words a meromorphic function f on C belongs to $\mathcal{L}(D)$ if firstly f is holomorphic except at those $p \in C$ for which $n_p > 0$ and there the order of the pole is at most n_p, and secondly f has a zero of order at least $-n_p$ at every $p \in C$ such that $n_p < 0$.

It is easy to check that $\mathcal{L}(D)$ is a complex vector space. We define

$$l(D) = \dim \mathcal{L}(D).$$

Note that $l(\kappa)$ is the dimension of the space of all holomorphic differentials on C for any canonical divisor κ, since if f is a meromorphic function and ω is a meromorphic differential then $(f) + (\omega) \geq 0$ if and only if $f\omega$ is a holomorphic differential on C.

We have another corollary to proposition 6.29.

Corollary 6.35 *If $\deg D < 0$ then $l(D) = 0$.*

Proof. If f is a meromorphic function on C, not identically zero, such that

$$(f) + D \geq 0$$

then

$$\deg D = \deg\left((f) + D\right) \geq 0.$$

$\square$

Lemma 6.36 *If $D \sim D'$ then $l(D) = l(D')$.*

Proof. We simply note that if

$$D = D' + (g)$$

where g is a meromorphic function on C, then

$$f \mapsto fg$$

defines an isomorphism between $\mathcal{L}(D)$ and $\mathcal{L}(D')$. $\square$

We can now state the Riemann-Roch theorem for nonsingular projective curves in $\mathbf{P}_2$.

Theorem 6.37 (Riemann-Roch) *If D is any divisor on a nonsingular projective curve C of genus g in $\mathbf{P}_2$ and κ is a canonical divisor on C, then*

$$l(D) - l(\kappa - D) = deg(D) + 1 - g.$$

We can immediately deduce the second promised alternative definition of the genus g of C.

Corollary 6.38 *The genus of a nonsingular projective curve C in $\mathbf{P}_2$ equals the dimension $l(\kappa)$ of the vector space of all holomorphic differentials on C.*

Proof. The Riemann-Roch theorem with $D = 0$ tells us that

$$l(0) - l(\kappa) = 1 - g.$$

But $l(0)$ is the dimension of the vector space $\mathcal{L}(0)$ of holomorphic functions on C, and every holomorphic function on C is constant by exercise 5.4 and theorem 7.12 below, so $l(0) = 1$. Hence

$$l(\kappa) = g$$

as required. $\square$

This corollary tells us in particular that there is a nonzero holomorphic differential on any nonsingular projective curve C of genus $g > 0$.

Another application of the Riemann-Roch theorem is the proof of the law of associativity for the additive group structure on a nonsingular cubic.

Theorem 6.39 *Let C be a nonsingular projective curve of degree 3 in $\mathbf{P}_2$ and let p_0 be a point of inflection on C. There is a unique additive group structure on C such that p_0 is the zero element and such that if $p, q, r \in C$ then*

$$p + q + r = 0$$

if and only if p, q and r are the three points of intersection (counting multiplicities) of C with a line in $\mathbf{P}_2$.

Proof. We have already proved this theorem except for the law of associativity on C (see theorem 3.38).

Let p, q, r be points of C, let

$$a = p + q,$$

$$b = a + r = (p + q) + r,$$

$$c = q + r$$

and
$$d = p + c = p + (q + r).$$

We wish to show that $b = d$. Since p, q and $-a$ are collinear there is a linear homogeneous polynomial in x, y, z which vanishes at p, q and $-a$. Similarly there is a linear homogeneous polynomial in x, y, z which vanishes at $a, -a$ and p_0. The ratio of these polynomials defines a meromorphic function ϕ on C with zeros at p and q and poles at a and p_0 (counting multiplicities). By the same argument there is a meromorphic function ψ on C with zeros at a and r and poles at b and p_0. Then $\phi\psi$ is a meromorphic function on C with zeros at p, q and r and poles at p_0 (with multiplicity two) and b. Similarly there is a meromorphic function on C with zeros at p, q and r and poles at p_0 (with multiplicity two) and d. If $b \neq d$ the ratio of these functions is a meromorphic function on C with a simple zero at d and a simple pole at b and no other zeros or poles.

By the degree-genus formula (§4.1) the genus of C is one, so if we regard the point b of C as a divisor and if κ is a canonical divisor it follows from proposition 6.31 and corollary 6.35 that

$$l(\kappa - b) = 0$$

so by the Riemann-Roch theorem

$$l(b) = deg(b) + 1 - g = 1.$$

This means that the only meromorphic functions on C with at most simple poles at b are the constant functions. Therefore $b = d$ as required. $\square$

Example 6.40 Let
$$H = \sum_{p \in C} I_p(C, L) p$$

where $I_p(C, L)$ is the intersection multiplicity at p of C with a line L in $\mathbf{P}_2$ defined by a linear homogeneous polynomial $R(x, y, z)$. Let us consider what the Riemann-Roch theorem means for a divisor of the form mH for some positive integer m.

By Bézout's theorem (3.1) H is a divisor of degree d on C where d is the degree of C. Thus
$$deg(\kappa - mH) = deg\,\kappa - md$$

which is strictly negative if m is large enough. Hence if m is large enough then
$$l(\kappa - mH) = 0$$

by corollary 6.35.

Now if $Q(x, y, z)$ is any homogeneous polynomial of degree m then

$$\frac{Q(x,y,z)}{R(x,y,z)^m}$$

defines a meromorphic function f on C such that

$$(f) + mH \geq 0,$$

i.e. an element of $\mathcal{L}(mH)$. Moreover two such polynomials define the same function on C if and only if their difference is divisible by the homogeneous polynomial $P(x, y, z)$ of degree d which defines C. Hence if

$$\mathbf{C}_k[x, y, z]$$

denotes the space of homogeneous polynomials of degree k in x, y, z then

$$
\begin{aligned}
l(mH) &\geq \dim \mathbf{C}_m[x,y,z]/P(x,y,z)\mathbf{C}_{m-d}[x,y,z] \\
&= \dim \mathbf{C}_m[x,y,z] - \dim \mathbf{C}_{m-d}[x,y,z] \\
&= \tfrac{1}{2}(m+1)(m+2) - \tfrac{1}{2}(m-d+1)(m-d+2) \\
&= md + \tfrac{1}{2}d(3-d) \\
&= md + 1 - g
\end{aligned}
$$

by the degree-genus formula (§4.1). Thus we have shown that

$$l(mH) - l(\kappa - mH) \geq \deg(mH) + 1 - g$$

when m is large enough. The Riemann-Roch theorem says that we have equality here, or equivalently that *every* meromorphic function f on C satisfying

$$(f) + mH \geq 0$$

is of the form

$$\frac{Q(x,y,z)}{R(x,y,z)^m}$$

for some $Q(x, y, z) \in \mathbf{C}_m[x, y, z]$, and in particular it is rational. The same argument shows that if $L_1, \ldots, L_m$ are lines in $\mathbf{P}_2$ and

$$H_j = \sum_{p \in C} I_p(C, L_j)\, p$$

for $1 \leq j \leq m$, then if m is large enough every meromorphic function f on C satisfying

$$(f) + H_1 + \cdots + H_m \geq 0$$

is rational. But it is easy to see that every meromorphic function f on C satisfies

$$(f) + H_1 + \cdots + H_m \geq 0$$

for some such $H_1, \ldots, H_m$. Thus we have proved that the following special case of Chow's theorem is a consequence of the Riemann-Roch theorem.

Theorem 6.41 *All meromorphic functions on a nonsingular projective curve in $\mathbf{P}_2$ are rational.*

In order to prove the Riemann-Roch theorem we need two lemmas whose proofs will be left to the end of this section.

Lemma 6.42 *Given any divisor D on C and any positive integer m_0 there exists $m \geq m_0$ and points $p_1, \ldots, p_k$ of C (not necessarily distinct) such that*

$$D + p_1 + \cdots + p_k \sim mH$$

where H is defined as in example 6.40.

Lemma 6.43 *If D is any divisor on C, κ is a canonical divisor and p is any point of C then*

$$0 \leq l(D + p) - l(\kappa - D - p) - l(D) + l(\kappa - D) \leq 1.$$

From these two lemmas we obtain half the Riemann-Roch theorem.

Corollary 6.44 (Riemann's theorem.) *If D is any divisor on C then*

$$l(D) - l(\kappa - D) \geq deg\, D - g + 1.$$

Proof. We saw in example 6.40 that there exists a positive integer m_0 such that if $m \geq m_0$ then

$$l(mH) - l(\kappa - mH) \geq deg(mH) - g + 1.$$

By lemma 6.42 we can choose $m \geq m_0$ and points $p_1, \ldots, p_k$ of C such that

$$D + p_1 + \cdots + p_k \sim mH.$$

Hence by corollary 6.30

$$deg(mH) = deg(D + p_1 + \cdots + p_k) = deg(D) + k,$$

and by lemma 6.36

$$l(mH) - l(\kappa - mH) = l(D + p_1 + \cdots + p_k) - l(\kappa - D - p_1 - \cdots - p_k)$$

Also by lemma 6.43 and induction on k we have

$$l(D + p_1 + \cdots + p_k) - l(\kappa - D - p_1 - \cdots - p_k) - l(D) + l(\kappa - D) \leq k.$$

Combining these inequalities and equalities we obtain

$$
\begin{aligned}
l(D) - l(\kappa - D) &\geq l(D + p_1 + \cdots + p_k) - l(\kappa - D - p_1 - \cdots - p_k) - k \\
&= l(mH) - l(\kappa - mH) - k \\
&\geq deg(mH) - g + 1 - k \\
&= deg\, D - g + 1
\end{aligned}
$$

as required. $\square$

From this we can prove the Riemann-Roch theorem, which says that in fact equality holds in 6.44.

Proof of theorem 6.37 (Riemann-Roch). Let D be any divisor on C and let κ be a canonical divisor. By proposition 6.31

$$deg\, \kappa = 2g - 2.$$

By corollary 6.44 applied first to D and then to $\kappa - D$ we have

$$l(D) - l(\kappa - D) \geq deg\, D - g + 1$$

and

$$
\begin{aligned}
l(\kappa - D) - l(D) &\geq deg(\kappa - D) - g + 1 \\
&= 2g - 2 - deg\, D - g + 1 \\
&= -deg\, D + g - 1.
\end{aligned}
$$

We deduce that equality holds, i.e. that the Riemann-Roch theorem is true. $\square$

It remains to prove propositions 6.29 and 6.31 and lemmas 6.42 and 6.43. The proof of lemma 6.42 is straightforward.

Proof of lemma 6.42. Let

$$D + \sum_{p \in C} n_p\, p$$

be any divisor on C and let m_0 be any positive integer. We wish to show that there exist points $p_1, \ldots, p_k \in C$ and $m \geq m_0$ such that

$$D + p_1 + \cdots + p_k \sim mH.$$

By adding points p to D we may assume without loss of generality that $n_p \geq 0$ for all $p \in C$ and that $deg\, D \geq m_0$.

For each of the finitely many $p \in C$ such that $n_p > 0$ we can choose a line in $\mathbf{P}_2$ through p whose points of intersection with C (allowing for multiplicities) are

$$q_1^{(p)} = p, q_2^{(p)}, \ldots, q_d^{(p)}.$$

Note that the ratio of two linear homogeneous polynomials in x, y, z defines a meromorphic function on C so if $q_1, \ldots, q_d$ are the points of intersection of any line in $\mathbf{P}_2$ with C (allowing for multiplicities) then

$$q_1 + \cdots + q_d \sim H.$$

Thus setting $m = deg\, D$ we have $m \geq m_0$ and

$$mH \sim \sum_{n_p > 0} n_p \sum_{1 \leq i \leq d} q_i^{(p)} = D + p_1 + \cdots + p_k$$

for suitable $p_1, \ldots, p_k$ where $k = m(d-1)$. □

The proof of proposition 6.31 is also fairly straightforward, given proposition 6.29.

Proof of proposition 6.31. By proposition 6.29 it is enough to show that there is some meromorphic differential ω on C, not identically zero, such that

$$deg(\omega) = 2g - 2.$$

Let $P(x, y, z)$ be a homogeneous polynomial of degree d defining C. We may assume that coordinates have been chosen so that $[0, 1, 0] \notin C$, i.e. the coefficient of y^d in $P(x, y, z)$ is nonzero. Then since $P(x, y, z)$ is irreducible and $\frac{\partial P}{\partial y}$ is not identically zero, by the weak form of Bézout's theorem (3.9) there are only finitely many points in C where $\frac{\partial P}{\partial y}$ vanishes. As $[0, 1, 0] \notin C$ either x or z is nonzero at each of these points. Therefore by making a projective transformation of the form

$$x \mapsto x, \quad y \mapsto y, \quad z \mapsto \alpha x + z$$

we can assume that if $[a, b, c] \in C$ and

$$\frac{\partial P}{\partial y}(a, b, c) = 0$$

then $c \neq 0$.

Now let ω be the differential $d(x/z)$ of the meromorphic function x/z on C. Near points $[a, b, c] \in C$ with $c \neq 0$ and

$$\frac{\partial P}{\partial y}(a, b, c) \neq 0$$

we can take x/z as a local holomorphic chart on C (see the proof of proposition 5.28) and hence ω has no zeros or poles at such points.

At any point $[a, b, c] \in C$ such that $c = 0$ we have $a \neq 0$ and

$$\frac{\partial P}{\partial y}(a, b, c) \neq 0$$

by our choice of coordinates, so

$$v = \frac{z}{x}$$

is a local holomorphic chart on C and

$$\omega = d\left(\frac{1}{v}\right) = -v^{-2}dv$$

has a pole of multiplicity two. Moreover the assumption that

$$\frac{\partial P}{\partial y}(a, b, c) \neq 0$$

whenever $[a, b, c] \in C$ and $c = 0$ tells us that the line

$$z = 0$$

is nowhere tangent to C, so by the corollary 3.25 to Bézout's theorem it meets C in precisely d points. These points contribute $-2d$ to the degree of the divisor (ω).

Finally consider the points $[a, b, c] \in C$ such that

$$\frac{\partial P}{\partial y}(a, b, c) = 0.$$

These points are precisely the ramification points of the mapping $\phi : C \to \mathbf{P}_1$ defined by $\phi[x, y, z] = [x, z]$ (see §4.2). At these points c is nonzero by our choice of coordinates, and hence so is $\frac{\partial P}{\partial x}(a, b, c)$, for otherwise by Euler's relation (lemma 2.32) $\frac{\partial P}{\partial z}(a, b, c)$ would also be zero, and this can't happen since C is nonsingular. Therefore

$$u = \frac{y}{z}$$

is a holomorphic chart on C near such points, and locally x/z is a holomorphic function $f(u)$ of u satisfying

$$P(f(u), u, 1) = 0.$$

Differentiating this identity m times shows that if $f^{(k)}(u_0) = 0$ for $1 \leq k < m$ then

$$f^{(m)}(u_0) = -\frac{\partial^m P}{\partial y^m}(u_0, f(u_0), 1)/\frac{\partial P}{\partial x}(u_0, f(u_0), 1).$$

It follows that the smallest positive integer m such that $f^{(m)}(u_0) \neq 0$ is equal to the smallest positive integer m such that

$$\frac{\partial^m P}{\partial y^m}(u_0, f(u_0), 1) \neq 0.$$

Since

$$\omega = d(f(u)) = f'(u)du$$

this tells us that the multiplicity of the zero of ω at a ramification point of $\phi : C \to \mathbf{P}_1$ is precisely one less than the ramification index of ϕ at the point (see definition 4.3). By lemma 4.7 and lemma 4.8 we can assume that our coordinates have been chosen such that there are exactly $d(d-1)$ ramification points and ω has a zero of multiplicity one at each one. Hence these points contribute $d(d-1)$ to the degree of the divisor (ω).

This shows that

$$deg(\omega) = d(d-1) - 2d = d(d-3).$$

It now follows from the degree-genus formula (§4.1) that

$$deg(\omega) = 2g - 2$$

as required. $\square$

In order to prove lemma 6.43 we need

Lemma 6.45 *Let ω be a meromorphic differential on C with precisely one pole. Then this pole is not a simple pole (i.e. its multiplicity is at least two).*

Remark 6.46 This lemma and proposition 6.29 can both be regarded as special cases of the *residue theorem* which says that if ω is a meromorphic differential on C (in fact more generally on any compact Riemann surface) with poles at $p_1, \ldots, p_t$ then

$$\sum_{i=1}^{t} res\{\omega, p_i\} = 0.$$

Here the residue $res\{\omega, p\}$ of a meromorphic differential $\omega = f\, dg$ at a point p is defined to be the ordinary residue

$$res\{(f \circ \phi_\alpha^{-1})(g \circ \phi_\alpha^{-1})', \phi_\alpha(p)\}$$

where $\phi : U_\alpha \to V_\alpha$ is a holomorphic chart on C such that $p \in U_\alpha$. It can be checked that this definition is independent of the choice of holomorphic chart $\phi : U_\alpha \to V_\alpha$.

The proofs given below of proposition 6.29 and lemma 6.45 can be modified without much difficulty to provide a proof of the residue theorem. Lemma 6.45 follows immediately from the residue theorem since the residue of a meromorphic differential at a simple pole is always nonzero. Proposition 6.29 follows by considering the meromorphic differential $\frac{dg}{g}$ where g is a meromorphic function on C which is not identically zero.

Proof of lemma 6.43 given lemma 6.45. Let D be any divisor on the nonsingular projective curve C in $\mathbf{P}_2$, let $\kappa = (\omega)$ be a canonical divisor and let p be any point of C. We need to show that

$$0 \leq l(D+p) - l(\kappa - D - p) - l(D) + l(\kappa - D) \leq 1$$

where $l(D)$ is defined as the dimension of the space $\mathcal{L}(D)$ of meromorphic functions f on C satisfying

$$(f) + D \geq 0$$

together with the zero function.

Suppose that

$$D = \sum_{q \in C} n_q\, q.$$

Then $\mathcal{L}(D) = \mathcal{L}(D+p)$ if there is no meromorphic function f on C such that

$$(f) + D + p \geq 0$$

with the inequality an equality at the point p, in the sense that f has a pole at p of multiplicity *exactly* $n_p + 1$ (if $n_p \geq 0$) or a zero at p of multiplicity exactly $-n_p - 1$ (if $n_p < 0$). Otherwise $\mathcal{L}(D)$ is a linear subspace of $\mathcal{L}(D+p)$ of codimension 1. In particular

$$0 \leq l(D+p) - l(D) \leq 1$$

and similarly

$$0 \leq l(\kappa - D) - l(\kappa - D - p) \leq 1.$$

Therefore it remains to show that we cannot have

$$l(D+p) - l(D) = 1$$

and

$$l(\kappa - D) - l(\kappa - D - p) = 1$$

simultaneously. If so there exist meromorphic functions f and g on C satisfying

$$(f) + D + p \geq 0$$

and

$$(g) + \kappa - D \geq 0$$

with both inequalities being equalities at the point p. Then $fg\omega$ is a meromorphic differential on C such that

$$(fg\omega) = (f) + (g) + \kappa \geq -p$$

with a pole of order *exactly* one at p. That is, $fg\omega$ is holomorphic except for a simple pole at p, which contradicts lemma 6.45. $\square$

Finally we must prove proposition 6.29 and lemma 6.45. For this we need a modification of lemma 4.21. Recall the definition of a triangulation (4.9).

Lemma 6.47 *Let $\{p_1, \ldots, p_r, q_1, \ldots, q_s\}$ be a set of $r+s$ points in $\mathbf{C}$ with $r \geq 3$. Then there is a triangulation (V, E, F) of $\mathbf{P}_1$ such that $V = \{p_1, \ldots, p_r\}$ and the set $\{q_1, \ldots, q_s\}$ is contained in the interior of a face, i.e. there is some $f : \triangle \to \mathbf{P}_1$ in F such that*

$$q_j \in f(\triangle^\circ)$$

for $1 \leq j \leq s$. We may also assume that ∞ is in the interior of a different face.

Proof. The proof is by induction on r, like that of lemma 4.21. First suppose $r = 3$.

We can assume by making a linear transformation that 0 does not belong to the set $\{p_1, \ldots, p_3, q_1, \ldots, q_s\}$, and that the arguments (taken in the range $[0, 2\pi]$) of the complex numbers $p_1, \ldots, p_3, q_1, \ldots, q_s$ are distinct. We can choose a real number R such that

$$R > \max\{|p_1|, \ldots, |p_3|, |q_1|, \ldots, |q_s|\},$$

and we can reorder p_1, p_2, p_3 if necessary so that

$$\arg(p_1) < \arg(p_2) < \arg(p_3).$$

Then if $\varepsilon > 0$ is sufficiently small we can find a piecewise-smooth path e_1 in $\mathbf{C}$ from p_1 to p_2 consisting of the straight line segment from p_1 to the point $R\exp(\varepsilon + arg(p_1))$, the arc of the circle radius R from $R\exp(\varepsilon + arg(p_1))$ to $R\exp(-\varepsilon + arg(p_2))$ and the straight line segment from $R\exp(-\varepsilon + arg(p_2))$ to p_2. Similarly we can find piecewise-smooth paths e_2 and e_3 from p_2 to p_3 and p_3 to p_1 (see figure 6.4). If $\varepsilon > 0$ is sufficiently small it is straightforward to check that this gives us a triangulation of $\mathbf{P}_1$ with the set of vertices $V = \{p_1, p_2, p_3\}$, the set of edges $E = \{e_1, e_2, e_3\}$ and with two faces, one of which contains $\{q_1, \ldots, q_s\}$ in its interior.

Now assume the inductive hypothesis that $r > 3$ and that we have a triangulation of $\mathbf{P}_1$ with vertices $p_1, \ldots, p_{r-1}$ such that the set $\{q_1, \ldots, q_s\}$ is contained in the interior of a face $f \in F$. We modify the proof of lemma 4.21 to complete the induction step. First assume that p_r also lies in the interior of the face f, i.e. that $p_r = f(s)$ for some $s \in \triangle^\circ$. By composing f with a homeomorphism of

$$\triangle = \{(x, y) \in \mathbf{R}^2 \; : \; x \geq 0, \;\; y \geq 0, \;\; x + y \leq 1\}$$

of the form

$$(x, y) \mapsto (x + y)^\alpha (x, y)$$

for some $\alpha > 0$ we can assume that none of the points $q_1, \ldots, q_s$ lie on the image under $f : \triangle \to \mathbf{P}_1$ of the straight line segment γ_0 from s to $(1, 0)$ in

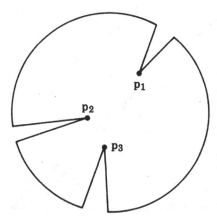

Figure 6.4: A triangulation of $\mathbf{P}_1$

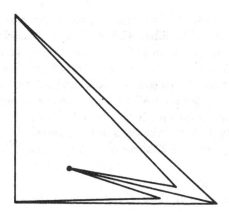

Figure 6.5: A subdivision of a triangle

$\triangle$. Then we choose points s_1 and s_2 of $\triangle^\circ$ close to $(1,0)$ and lying above and below the straight line segment γ_0. Let γ_1 be the join of the straight line segments from s to s_1 and from s_1 to $(0,1)$ and let γ_2 be the join of the straight line segments from s to s_2 and from s_2 to $(0,0)$ (see figure 6.5). If s_1 and s_2 are sufficiently close to $(1,0)$ then it is straightforward to check that there is a triangulation of $\mathbf{P}_1$ with vertices $p_1,\ldots,p_r$ and edges those of the original triangulation together with $f \circ \gamma_0, f \circ \gamma_1$ and $f \circ \gamma_2$, such that the points $q_1,\ldots,q_s$ all lie in the interior of some face.

If p_r lies in the interior of a different face from $q_1,\ldots,q_s$ then we can complete the induction step exactly as in the proof of lemma 4.21. Finally if p_r lies on an edge then we can complete the induction step by modifying the proof of lemma 4.21 along the lines described above (see figure 6.6). $\square$

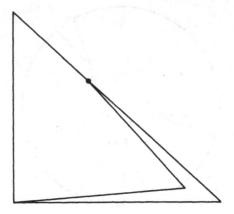

Figure 6.6: Another subdivision of a triangle

Proof of lemma 6.45. Let us suppose, to obtain a contradiction, that $\omega = g\, dh$ is a meromorphic differential on C with precisely one pole which is a simple pole, at q, say. As in §4.2 we choose coordinates in $\mathbf{P}_2$ such that $[0, 1, 0] \notin C$ and define $\phi : C \to \mathbf{P}_1$ by $\phi[x, y, z] = [x, z]$. We may also assume that the coordinates are chosen so that $0, \phi(q)$ and ∞ are distinct and are not branch points of ϕ. By lemma 6.47 we may choose a triangulation (V, E, F) of $\mathbf{P}_1$ such that every branch point of ϕ belongs to V, and $\phi(q)$ and 0 are in the interior of a face $f_0 \in F$ while ∞ is in the interior of a different face f_∞. By the proof of proposition 4.22 there is a triangulation $(\tilde{V}, \tilde{E}, \tilde{F})$ of C given by

$$\begin{aligned}
\tilde{V} &= \phi^{-1}(V), \\
\tilde{E} &= \{\tilde{e} : [0, 1] \to C \text{ continuous} : \phi \circ \tilde{e} \in E\}
\end{aligned}$$

and

$$\tilde{F} = \{\tilde{f} : \Delta \to C \text{ continuous} : \phi \circ \tilde{f} \in F\}.$$

By subdividing the triangulation if necessary we can assume that each face has at most one branch point among its vertices. This means that if $\tilde{f} \in \tilde{F}$ and $f = \phi \circ \tilde{f} \in F$ then by c.6, c.7, c.8 and c.9 of appendix C

$$\phi : \tilde{f}(\Delta) \to f(\Delta)$$

is a homeomorphism whose restriction to $\tilde{f}(\Delta - \{(0, 0), (1, 0), (0, 1)\})$ of C is the restriction of a holomorphic chart on C if $f \neq f_\infty$ (see the proof of proposition 5.28); if $f = f_\infty$ we must compose ϕ with the mapping $z \mapsto \frac{1}{z}$. The boundary of $\tilde{f}(\Delta)$ in C is the image of the join $\tilde{\gamma}$ of the paths $\tilde{f} \circ \sigma_i$ for $1 \leq i \leq 3$ where $\sigma_1, \sigma_2, \sigma_3 : [0, 1] \to \Delta$ are defined as in 4.9(iii). The composition $\gamma = \phi \circ \tilde{\gamma}$ of this closed piecewise-smooth path in C with ϕ is a closed piecewise-smooth path in $\mathbf{P}_1$ whose image is the boundary of $f(\Delta)$, and by definition 6.15

$$\int_{\tilde{\gamma}} \omega = \int_{\gamma} (\phi \, |_{\bar{f}(\Delta)}^{-1})^{*} \omega$$

$$= \int_{\gamma} (g \circ \phi \, |_{\bar{f}(\Delta)}^{-1})(h \circ \phi \, |_{\bar{f}(\Delta)}^{-1})'(z) dz.$$

If $q \in \tilde{f}(\Delta^{\circ})$ then $f = f_0$ and

$$(g \circ \phi \, |_{\bar{f}(\Delta)}^{-1})(h \circ \phi \, |_{\bar{f}(\Delta)}^{-1})'$$

has a simple pole at $\phi(q)$ inside γ and no other poles. Since the residue of a meromorphic function at a simple pole in $\mathbf{C}$ is always nonzero it follows from Cauchy's residue theorem that

$$\int_{\tilde{\gamma}} \omega \neq 0$$

if $q \in \tilde{f}(\Delta^{\circ})$. On the other hand if $q \notin \tilde{f}(\Delta^{\circ})$ then

$$(g \circ \phi \, |_{\bar{f}(\Delta)}^{-1})(h \circ \phi \, |_{\bar{f}(\Delta)}^{-1})'$$

has no poles inside γ and so

$$\int_{\tilde{\gamma}} \omega = 0.$$

Since q belongs to $\tilde{f}(\Delta^{\circ})$ for precisely one $\tilde{f} \in \tilde{F}$, if we sum over all $\tilde{f} \in \tilde{F}$ we get

$$\sum_{\tilde{f} \in \tilde{F}} \int_{\tilde{\gamma}} \omega \neq 0.$$

On the other hand each integral $\int_{\tilde{\gamma}} \omega$ can be expressed as the sum of the integrals of ω along the three edges e_f^1, e_f^2, e_f^3 with signs $\pm$ depending on whether $e_f^i = f \circ \sigma_i$ or $e_f^i = f \circ \sigma_i \circ r$ for $1 \leq i \leq 3$ (see definition 4.9(iii)). Moreover by 4.9(v) for every $e \in E$ there is exactly one face $f_e^+ \in F$ such that $e = f_e^+ \circ \sigma_i$ and exactly one face $f_e^- \in F$ such that $e = f_e^- \circ \sigma_i \circ r$. Thus the integrals of ω along edges occurring in the sum

$$\sum_{\tilde{f} \in \tilde{F}} \int_{\tilde{\gamma}} \omega$$

cancel in pairs and so we get

$$\sum_{\tilde{f} \in \tilde{F}} \int_{\tilde{\gamma}} \omega = 0.$$

This contradiction proves lemma 6.45. □

Proof of proposition 6.29. Let g be a meromorphic function on C which is not identically zero. We need to show that g has the same number of zeros as poles, counting multiplicities. We consider the meromorphic differential dg/g on C, which has poles precisely at the points $q_1, \ldots, q_t$ where g has zeros or poles.

As in the proof of lemma 6.45 we choose coordinates in $\mathbf{P}_2$ such that $[0, 1, 0] \notin C$ and $0, \phi(q_1), \ldots, \phi(q_t)$ and ∞ are distinct and are not branch points of the map $\phi : C \to \mathbf{P}_1$ defined by $\phi[x, y, z] = [x, z]$. Using lemma 6.47 we choose a triangulation (V, E, F) of $\mathbf{P}_1$ such that every branch point of ϕ belongs to V, and $\phi(q_1), \ldots, \phi(q_t)$ and 0 are in the interior of a face $f_0 \in F$ while ∞ is in the interior of a different face $f_\infty \in F$. As in the proof of lemma 6.45 we get a triangulation $(\tilde{V}, \tilde{E}, \tilde{F})$ of C such that

$$\sum_{\tilde{f} \in \tilde{F}} \int_{\tilde{\gamma}} \frac{dg}{g} = 0$$

where $\tilde{\gamma}$ is the join of the paths $\tilde{f} \circ \sigma_i$ for $1 \le i \le 3$ (see definition 4.9(iii)). Also

$$\int_{\tilde{\gamma}} \frac{dg}{g} = \int_\gamma \frac{(g \circ \phi \mid_{\tilde{f}(\Delta)}^{-1})'(z)}{(g \circ \phi \mid_{\tilde{f}(\Delta)}^{-1})(z)} dz$$

for each $\tilde{f} \in \tilde{F}$ where $\gamma = \phi \circ \tilde{\gamma}$ is a closed piecewise-smooth path in $\mathbf{C}$. (If $\phi \circ \tilde{f} = f_\infty$ we must compose with the function $z \mapsto \frac{1}{z}$ here). By remark 5.1 the meromorphic function

$$\frac{(g \circ \phi \mid_{\tilde{f}(\Delta)}^{-1})'(z)}{(g \circ \phi \mid_{\tilde{f}(\Delta)}^{-1})(z)}$$

has a pole at a point a inside γ with residue ρ if and only if $g \circ \phi \mid_{\tilde{f}(\Delta)}^{-1}$ has either a zero at a with multiplicity ρ or a pole at a with multiplicity $-\rho$. Since the restriction of ϕ to $\tilde{f}(\Delta^\circ)$ is a holomorphic chart on C the latter statement is equivalent to saying that g has either a zero at $\phi \mid_{\tilde{f}(\Delta)}^{-1}(a)$ with multiplicity ρ or a pole at $\phi \mid_{\tilde{f}(\Delta)}^{-1}(a)$ with multiplicity $-\rho$. Therefore by Cauchy's residue theorem

$$\int_{\tilde{\gamma}} \frac{dg}{g} = \pm(Z(\tilde{f}) - P(\tilde{f}))$$

where $Z(\tilde{f})$ and $P(\tilde{f})$ are the numbers of zeros and poles (counted with multiplicities) of g in the interior of the face $\tilde{f}$, and where the sign depends on whether γ is positively or negatively oriented in $\mathbf{C}$. Since all the zeros and poles of g lie in $\phi^{-1}(f_0(\Delta^\circ))$ and the sign is consistent for those $\tilde{f} \in \tilde{F}$ such that $\phi \circ \tilde{f} = f_0$, summing over all such $\tilde{f}$ gives

$$\sum_{\tilde{f} \in \tilde{F}, \phi \circ \tilde{f} = f_0} \int_{\tilde{\gamma}} \frac{dg}{g} = \pm(Z - P)$$

where Z and P are the total numbers of zeros and poles counted with multiplicities of g in C, while summing over the other $\tilde{f} \in \tilde{F}$ gives

$$\sum_{\tilde{f} \in \tilde{F}, \phi \circ \tilde{f} \neq f_0} \int_{\tilde{\gamma}} \frac{dg}{g} = 0.$$

Combining these results we get

$$Z - P = 0$$

as required. $\square$

6.4 Exercises

6.1. A holomorphic differential of the form dg where g is a holomorphic function on a Riemann surface S is called *exact*. Show that the integral of an exact holomorphic differential along a closed piecewise-smooth path in S is zero. Deduce that the holomorphic differential η on a torus $\mathbf{C}/\Lambda$ satisfying $\pi^* \eta = dz$ (see the discussion before proposition 6.17) is not exact.

6.2. Two piecewise-smooth closed paths γ_0 and $\gamma_1 : [0,1] \to S$ in a Riemann surface S are called *homotopic* if there is a piecewise-smooth map

$$G : [0,1] \times [0,1] \to S$$

with $G(0,t) = \gamma_0(t), G(1,t) = \gamma_1(t)$ and $G(t,0) = G(t,1)$ for all $t \in [0,1]$. The deformation theorem from complex analysis (see e.g. [Priestley 85] 4.10) says that if γ_0 and γ_1 are homotopic closed piecewise-smooth paths in an open subset U of $\mathbf{C}$ and if $f : U \to \mathbf{C}$ is holomorphic then

$$\int_{\gamma_0} f(z)\, dz = \int_{\gamma_1} f(z)\, dz.$$

Deduce that if γ_0 and γ_1 are homotopic closed piecewise-smooth paths in a Riemann surface S and η is a holomorphic differential on S then

$$\int_{\gamma_0} \eta = \int_{\gamma_1} \eta.$$

[Hint: show that if $G : [0,1] \times [0,1] \to S$ is continuous then there is a positive integer N such that if $1 \leq j, k \leq N$ then there is a holomorphic chart $\phi_{jk} : U_{jk} \to V_{jk}$ on an open subset U_{jk} of S with the $\frac{1}{N}$ by $\frac{1}{N}$ square

$$D_{jk} = [(j-1)/N, j/N] \times [(k-1)/N, k/N] \subseteq G^{-1}(U_{jk}).$$

Use the deformation theorem from complex analysis to show that the integral of η around the path in S given by restricting G to the boundary of the square D_{jk} is zero. Add up all these integrals to obtain the result.]

6.3. The holomorphic differential dz on $\mathbf{C}$ extends to a meromorphic differential dz on $\mathbf{P}_1 = \mathbf{C} \cup \{\infty\}$. Show that this differential has a pole of order 2 at ∞. Show that every meromorphic differential on $\mathbf{P}_1$ can be written as $f\,dz$ where f is a meromorphic function on $\mathbf{P}_1$, and that $f\,dz$ is meromorphic if and only if f is holomorphic on $\mathbf{C}$ and $z^2 f(z)$ tends to a finite limit as $z \to \infty$. Deduce that there are no holomorphic differentials on $\mathbf{P}_1$.

6.4. Complete the proof of lemma 6.25 by showing that the map μ from $\mathbf{P}_1$ to $\mathbf{C}/\Lambda$ is continuous at any $[b, c] \in \mathbf{P}_1$.
[Hint: this should be obvious. To prove it rigorously it is enough to show that every sequence $[b_n, c_n]$ in $\mathbf{P}_1$ converging to $[b, c]$ has a subsequence $[b_{n_r}, c_{n_r}]$ such that the points $p_i(b_{n_r}, c_{n_r})$ in C_Λ converge to $p_i(b, c)$ for $i = 1, 2, 3$, where $p_1(b, c), p_2(b, c)$ and $p_3(b, c)$ can be reordered if necessary. To prove this, note that since C_Λ is compact there is a subsequence $[b_{n_r}, c_{n_r}]$ such that $p(b_{n_r}, c_{n_r}), q(b_{n_r}, c_{n_r})$ and $r(b_{n_r}, c_{n_r})$ converge to some points p, q, r in C_Λ. Show that

$$\{p, q, r\} \subseteq \{p_1(b, c), p_2(b, c), p_3(b, c)\}.$$

If $p_1(b, c) = p_2(b, c) = p_3(b, c)$ then clearly

$$\{p, q, r\} = \{p_1(b, c), p_2(b, c), p_3(b, c)\}.$$

If $p_1(b, c), p_2(b, c)$ and $p_3(b, c)$ are distinct then the proof of lemma 6.25 shows that

$$\{p, q, r\} = \{p(b, c), q(b, c), r(b, c)\}.$$

Modify the proof of lemma 6.25 to show that the same is true in the remaining case.]

6.5. Let C_Λ be the nonsingular cubic curve in $\mathbf{P}_2$ associated with a lattice Λ in $\mathbf{C}$. Show that if $a_1 \neq a_2$ and $p = [a_1, b_1, 1], q = [a_2, b_2, 1]$ are points of C_Λ then the line through p and q in $\mathbf{P}_2$ meets C_Λ again at $r = [a_3, b_3, 1]$ where

$$a_3 = \frac{1}{4}\left(\frac{b_1 - b_2}{a_1 - a_2}\right)^2 - (a_1 + a_2),$$

$$b_3 = \left(\frac{b_1 - b_2}{a_1 - a_2}\right)a_3 + \left(\frac{a_1 b_2 - a_2 b_1}{a_1 - a_2}\right).$$

Use Abel's theorem to deduce that

$$\wp(z_1 + z_2) = \frac{1}{4}\left(\frac{\wp'(z_1) - \wp'(z_2)}{\wp(z_1) - \wp(z_2)}\right)^2 - \wp(z_1) - \wp(z_2)$$

if $z_1 \notin \Lambda \pm z_2$ and $z_1, z_2 \notin \Lambda$. Hence show that

$$\wp(2z) = \frac{1}{4}\left(\frac{\wp''(z)}{\wp'(z)}\right)^2 - 2\wp(z).$$

6.6. Show that a point $p \neq [0, 1, 0]$ of the cubic curve C_Λ associated with a lattice Λ in $\mathbf{C}$ has order two and only if the tangent to C_Λ at p passes through $[0, 1, 0]$. Show that the points of order one or two form a subgroup of C_Λ isomorphic to $C_2 \times C_2$.

6.7. Deduce from Abel's theorem that if n is a positive integer there are exactly n^2 elements of C_Λ with order dividing n, and they form a subgroup of C_Λ isomorphic to the product of two cyclic groups of order n.

By considering the equation $p + p = -q$ or otherwise, show that if $q \in C_\Lambda$ is not a point of inflection then there are exactly four points in C_Λ other than q itself whose tangent lines pass through q.

6.8. Use Abel's theorem to show that if $u, v, w \in \mathbf{C} - \Lambda$ and u, v, w are distinct modulo Λ then $u + v + w \in \Lambda$ if and only if

$$0 = \det\begin{pmatrix} \wp(u) & \wp'(u) & 1 \\ \wp(v) & \wp'(v) & 1 \\ \wp(w) & \wp'(w) & 1 \end{pmatrix}.$$

6.9. (For topologists only). Show that if C is a nonsingular projective curve with a group structure such that the group multiplication defines a continuous map from $C \times C$ to C then $\chi(C) = 0$ so C has genus 1 and degree 3. [Hint: use the Lefschetz fixed point theorem, which implies that if $\chi(C) \neq 0$ then any continuous map $h : C \to C$ homotopic to the identity has a fixed point. Consider the map $p \mapsto p + q$ for fixed $q \in C$.]

6.10. Use remark 6.22 and exercise 2.10 to show that there is a projective transformation of $\mathbf{P}_2$ taking the points of inflection of C_Λ to the points

$$\begin{array}{ccc} [0, 1, -1], & [-1, 0, 1], & [1, -1, 0], \\ [0, 1, \alpha], & [\alpha, 0, 1], & [1, \alpha, 0], \\ [0, \alpha, 1], & [1, 0, \alpha], & [\alpha, 1, 0], \end{array}$$

where $\alpha^2 - \alpha + 1 = 0$, and taking C_Λ to the curve defined by

$$x^3 + y^3 + z^3 + 3\mu xyz = 0$$

for some $\mu \in \mathbf{C}$.

6.11. Let S be a compact connected Riemann surface of genus zero. Assuming that the Riemann-Roch theorem applies to S, show that if a divisor D on S is of the form $D = p$ for any $p \in S$ then

$$l(D) = 2.$$

Deduce that there is a meromorphic function f on S with a simple pole at p and no other poles. Show that $f : S \to \mathbf{P}_1$ is a holomorphic bijection with holomorphic inverse.

6.12. The *periods*

$$\pi_1(\lambda) = 2 \int_0^1 \frac{dx}{\sqrt{x(x-1)(x-\lambda)}}$$

and

$$\pi_2(\lambda) = 2 \int_1^\lambda \frac{dx}{\sqrt{x(x-1)(x-\lambda)}}$$

of the cubic curve defined by

$$y^2 z = z(x - z)(x - \lambda z)$$

are holomorphic functions of $\lambda \in \mathbf{C} - \{0, 1\}$. The sum

$$\pi(\lambda) = \pi_1(\lambda) + \pi_2(\lambda)$$

has a removable singularity at $\lambda = 0$ with $\pi(0) = 2\pi$. Show that π_1, π_2 and π satisfy the *Picard-Fuchs equation*

$$0 = \frac{1}{4}\pi + (2\lambda - 1)\frac{d\pi}{d\lambda} + \lambda(\lambda - 1)\frac{d^2\pi}{d\lambda^2}.$$

Deduce that

$$\pi(\lambda) = 2\pi \sum_{n=0}^{\infty} a_n \lambda^n$$

where $a_0 = 1$ and

$$(n + 1)^2 a_{n+1} = (n + \frac{1}{2})^2 a_n$$

so that

$$\pi(\lambda) = 2\pi \sum_{n=0}^{\infty} \binom{2n}{n}^2 2^{-4n} \lambda^n.$$

6.13. Let C be a nonsingular projective curve in $\mathbf{P}_2$. By applying a suitable projective transformation we can assume that C is defined by

$$y^2 z = x(x - z)(x - \lambda z)$$

for some $\lambda \in \mathbf{C} - \{0,1\}$. The following steps (which are not to be done in too much detail, if at all!) give a proof that C is biholomorphic to a torus $\mathbf{C}/\Lambda$. Assume first that $\lambda \notin \mathbf{R}$.

(i). Show that there is a holomorphic function h from $\mathbf{C} - \{[0,1,0]\}$ to $\mathbf{P}_1$ defined by $h[x,y,z] = [x,z]$.

(ii). If $a \neq b$ are complex numbers let L_{ab} be the (real) line in $\mathbf{C}$ through a and b. Let $V_{\lambda,+}$ and $V_{\lambda,-}$ (respectively $V_{0,\pm}, V_{1,\pm}$) be the connected components of $\mathbf{C} - L_{01}$ (respectively $\mathbf{C} - L_{0\lambda}, \mathbf{C} - L_{1\lambda}$) labelled so that $\lambda \in V_{\lambda,+}$ (and $0 \in V_{0,+}, 1 \in V_{1,+}$). Let $U_\lambda = V_{\lambda,-} \cup (V_{0,+} \cap V_{1,+})$ and define U_0 and U_1 similarly.

Show that $\mathbf{P}_1 - \{0,1,\lambda\} = U_0 \cup U_1 \cup U_\lambda$. Show that if $\xi \in \{0,1,\lambda\}$ then the inverse image $h^{-1}(U_\xi)$ of U_ξ in C has two connected components $W_{\xi,\pm}$, and use the implicit function theorem to show that the restriction of h to each of these is a biholomorphism with inverse $g_{\xi,\pm} : U_\xi \to W_{\xi,\pm}$ of the form

$$w \to [w, \pm f_\xi(w), 1]$$

where $f_\xi : U_\xi \to \mathbf{C}$ is holomorphic and satisfies

$$f_\xi(w)^2 = w(w-1)(w-\lambda).$$

(iii). If a and b are distinct complex numbers define

$$\gamma_{ba} : [0,1] \to \mathbf{C} \quad \text{and} \quad \gamma_{\infty a} : [0,1] \to \mathbf{C}$$

by

$$\gamma_{ba}(t) = b + t(a-b) \quad \text{and} \quad \gamma_{\infty a}(t) = a/t.$$

For $\xi, \zeta \in \{0,1,\lambda\}, \xi \neq \zeta$, define

$$\phi_{\xi\zeta} : W_\xi = W_{\xi,+} \cup W_{\xi,-} \to \mathbf{C}$$

by

$$\phi_{\xi\zeta}[a,b,1] = \int_{g_{\xi,\pm}\circ\gamma_{\zeta a}} y^{-1}\, dx = \pm \int_{\gamma_{\zeta a}} (f_\xi(z))^{-1}\, dz$$

where the sign depends on whether $[a,b,1]$ belongs to $W_{\xi,+}$ or $W_{\xi,-}$. Show that $\phi_{\xi\zeta}$ is holomorphic, and that $\phi_{\xi\zeta}[a,-b,1] = -\phi_{\xi\zeta}[a,b,1]$.

(iv). Let $\omega_0 = 2\int_{\gamma_{1\lambda}} f_0(z)^{-1}\, dz$ and $\omega_1 = 2\int_{\gamma_{\lambda 0}} f_1(z)^{-1}\, dz$. Use Cauchy's theorem to show that

$$\begin{array}{ll}
\phi_{01}[a,b,1] - \phi_{0\lambda}[a,b,1] = \pm\frac{1}{2}\omega_0 & \text{if } [a,b,1] \in W_0, \\
\phi_{1\lambda}[a,b,1] - \phi_{10}[a,b,1] = \pm\frac{1}{2}\omega_1 & \text{if } [a,b,1] \in W_1, \\
\phi_{\lambda 0}[a,b,1] - \phi_{\lambda 1}[a,b,1] = \pm\frac{1}{2}(\omega_0+\omega_1) & \text{if } [a,b,1] \in W_\lambda.
\end{array}$$

(v). Using (iv) show that there is a well-defined holomorphic map

$$\phi : W_0 \cup W_1 \cup W_\lambda = C - \{[0,0,1],[1,0,1],[\lambda,0,1],[0,1,0]\} \to \mathbf{C}/\Lambda,$$

where $\Lambda = \{n\omega_0 + m\omega_1 : n, m \in \mathbf{Z}\}$, such that

$$\phi(x) = \Lambda + \phi_{01}(x) + \tfrac{1}{2}\omega_1 = \Lambda + \phi_{0\lambda}(x) + \tfrac{1}{2}(\omega_0 + \omega_1) \quad \text{if } x \in W_0,$$
$$\phi(x) = \Lambda + \phi_{1\lambda}(x) + \tfrac{1}{2}(\omega_0 + \omega_1) = \Lambda + \phi_{10}(x) + \tfrac{1}{2}\omega_0 \quad \text{if } x \in W_1,$$
$$\phi(x) = \Lambda + \phi_{\lambda 0}(x) + \tfrac{1}{2}\omega_0 = \Lambda + \phi_{\lambda 1}(x) + \tfrac{1}{2}\omega_1 \qquad \text{if } x \in W_\lambda.$$

Show that $\phi[a, -b, 1] = -\phi[a, b, 1]$.
(vi). Show that ϕ extends to a continuous map $\phi : C \to \mathbf{C}/\Lambda$ such that

$$\phi[0,0,1] = \Lambda + \tfrac{1}{2}\omega_0, \qquad \phi[1,0,1] = \Lambda + \tfrac{1}{2}\omega_1,$$
$$\phi[\lambda,0,1] = \Lambda + \tfrac{1}{2}(\omega_0 + \omega_1), \quad \phi[0,1,0] = \Lambda + 0.$$

Use theorem B.6 of Appendix B to show that ϕ is holomorphic.
(vii). Show that $\phi : C \to \mathbf{C}/\Lambda$ is a holomorphic bijection and use exercise 5.6 to deduce that ϕ has a holomorphic inverse.
(viii). Show that

$$\Lambda = \{ \int_\gamma y^{-1}\, dx \ : \ \gamma \text{ piecewise} - \text{smooth closed path in } \mathbf{C}\}.$$

(ix). Adapt this argument to apply to the case $\lambda \in \mathbf{R}$ by modifying the definitions of U_0, U_1 and U_λ.

6.14. The following steps build on exercise 6.13 to show that every nonsingular cubic curve C is equivalent under a projective transformation to one of the form C_Λ for some lattice Λ in $\mathbf{C}$. As before we may assume that C is defined in appropriate coordinates by

$$y^2 z = x(x - z)(x - \lambda z)$$

for some $\lambda \in \mathbf{C} - \{0, 1\}$. Exercise 6.13 shows that there is a lattice $\Lambda = \{n\omega_0 + m\omega_1 : n, m \in \mathbf{Z}\}$ in $\mathbf{C}$ and a holomorphic map $\phi : C \to \mathbf{C}/\Lambda$ with holomorphic inverse such that

$$\phi[0,0,1] = \Lambda + \tfrac{1}{2}\omega_0, \qquad \phi[1,0,1] = \Lambda + \tfrac{1}{2}\omega_1,$$
$$\phi[\lambda,0,1] = \Lambda + \tfrac{1}{2}(\omega_0 + \omega_1), \quad \phi[0,1,0] = \Lambda + 0,$$

and such that $\phi[a, -b, 1] = -\phi[a, b, 1]$.
(i). Show that the nonsingular cubic curve C_Λ defined by

$$y^2 z = 4x^3 - g_2(\Lambda)xz^2 - g_3(\Lambda)z^3$$

is equivalent under a projective transformation of the form

$$[x, y, z] \mapsto [ax + bz, y, z]$$

to the curve D defined by

$$y^2z = x(x-z)(x-\mu z)$$

for some $\mu \in \mathbf{C} - \{0,1\}$. Show that the composition of $\phi : C \to \mathbf{C}/\Lambda$ with the holomorphic bijection $u : \mathbf{C}/\Lambda \to C_\Lambda$ defined at definition 5.22 and this projective transformation is a holomorphic bijection $h : C \to D$ and we can choose the projective transformation such that

$$h[0,0,1] = [0,0,1], \qquad h[1,0,1] = [1,0,1]$$
$$h[\lambda,0,1] = [\mu,0,1] \quad \text{and} \quad h[0,1,0] = [0,1,0].$$

(ii). Use the fact that $\phi[a,-b,1] = -\phi[a,b,1]$ for all $[a,b,1] \in C$ and that $\wp$ is an even function to show that the function $h : C \to D$ of (i) restricts to a holomorphic bijection

$$h : C - \{[0,0,1],[0,1,0],[1,0,1],[\lambda,0,1]\} \to$$

$$D - \{[0,0,1],[0,1,0],[1,0,1],[\mu,0,1]\}$$

of the form

$$h[x,y,1] = [f(x),g(x,y),1]$$

where the first coordinate $f(x)$ depends *only* on x. Show that

$$f : \mathbf{C} - \{0,1,\lambda\} \to \mathbf{C} - \{0,1,\mu\}$$

is a holomorphic bijection such that

$$f(x) \to \begin{cases} 0 & \text{as } x \to 0, \\ 1 & \text{as } x \to 1, \\ \mu & \text{as } x \to \lambda. \end{cases}$$

Show that $0,1$ and λ are removable singularities of f and deduce that f extends to a holomorphic $f : \mathbf{C} \to \mathbf{C}$ such that $f(0) = 0, f(1) = 1$ and $f(\lambda) = \mu$.

(iii). Show that a holomorphic bijection $f : \mathbf{C} \to \mathbf{C}$ must be of the form $f(z) = az + b$ for some $a, b \in \mathbf{C}, a \neq 0$.

[One method of proof runs along the following lines. If the function $f(\frac{1}{z})$ has a pole at $z = 0$ then show that f must be a polynomial, and since $f : \mathbf{C} \to \mathbf{C}$ is a bijection it must then be a linear polynomial. Otherwise using the Casorati-Weierstrass theorem (see e.g. [Priestley 85] pp.87 and 102) we can find $w \in \mathbf{C}$ and a sequence $(a_n)_{n \geq 1}$ of complex numbers such that $a_n \to \infty$ and $f(a_n) \to w$ as $n \to \infty$, from which a contradiction can be obtained by showing that $a_n \to f^{-1}(w)$ as $n \to \infty$.]

(vi). Deduce from (iii) that the holomorphic bijection $f : \mathbf{C} \to \mathbf{C}$ of (ii) must be the identity map and hence $\lambda = \mu$. Conclude that C is equivalent under a projective transformation to the cubic curve C_Λ associated to the lattice Λ.

6.15. Use exercise 5.4 to show that any two nonzero holomorphic differentials on a compact connected Riemann surface are constant multiples of each other. Use exercises 5.15 and 6.14 to deduce that there is a bijection given by

$$\Lambda \mapsto C_\Lambda$$

between the set of equivalence classes of lattices Λ in $\mathbf{C}$ under the equivalence relation $\sim$ defined by $\Lambda \sim \tilde{\Lambda}$ if $\Lambda = a\tilde{\Lambda}$ for some $a \in \mathbf{C} - \{0\}$ and the set of equivalence classes of nonsingular cubic curves in $\mathbf{P}_2$ under projective transformations. Use exercise 6.13 to show that the inverse of this bijection is given by

$$C \mapsto \{ \int_\gamma \eta \; : \; \gamma \text{ closed piecewise} - \text{smooth path in } \mathbf{C} \}$$

where η is any nonzero holomorphic differential on C.

Chapter 7

Singular curves

Up to this point our attention has been given almost entirely to *nonsingular* curves. In this chapter we shall look at *singular* projective curves in $\mathbf{P}_2$. However the study of the singularities of curves is a huge subject (see e.g. [Brieskorn & Knörrer 86] and the references therein) and we shall be able to cover only a tiny fraction of it.

In §7.1 we shall construct a "resolution of singularities" $\tilde{C}$ for any singular projective curve C in $\mathbf{P}_2$, in the sense that $\tilde{C}$ is a compact Riemann surface with a surjective continuous map $\pi : \tilde{C} \to C$ such that $\pi^{-1}(Sing(C))$ is a finite set of points in $\tilde{C}$ and the restriction of π to the complement of this set in $\tilde{C}$ is a holomorphic bijection onto the nonsingular part of C. In §7.2 we shall investigate what a singular curve looks like near a singular point, using a method which goes back to Newton. Finally in §7.3 we shall give a description of the topology of singular curves.

7.1   Resolution of singularities

In this section we shall relate the complex projective curve

$$C = \{[x,y,z] \in \mathbf{P}_2 \ : \ P(x,y,z) = 0\}$$

to the holomorphic functions $y(x)$ defined on open subsets of $\mathbf{C}$ satisfying

$$P(x, y(x), 1) = 0.$$

The implicit function theorem of complex analysis (see Appendix B) tells us that if $[a, b, c] \in C$ satisfies

$$c \neq 0 \neq \frac{\partial P}{\partial y}(a, b, c)$$

185

then there is a holomorphic function $y(x)$ defined in a neighbourhood U of a/c in $\mathbf{C}$ such that

$$y(a/c) = b/c$$

and

$$P(x, y(x), 1) = 0$$

for all $x \in U$. We shall show how to associate a Riemann surface to any holomorphic function on a nonempty open subset of $\mathbf{C}$. We shall see that if C is irreducible then the Riemann surface associated to any function $y(x)$ satisfying

$$P(x, y(x), 1) = 0$$

is a resolution of singularities of C in the sense described above.

Definition 7.1 *In this section we shall call an ordered pair (f, g) of mero-morphic functions defined on an open neighbourhood of 0 in $\mathbf{C}$ simply a* pair *if f is not constant on any neighbourhood of 0 and the mapping defined by*

$$t \mapsto (f(t), g(t))$$

is one-to-one near 0. A parameter change *is a holomorphic function ρ defined on an open neighbourhood of 0 in $\mathbf{C}$ such that $\rho(0) = 0$ and $\rho'(0) \neq 0$. We say that two pairs (f, g) and $(\tilde{f}, \tilde{g})$ are* equivalent *and write*

$$(f, g) \sim (\tilde{f}, \tilde{g})$$

if there is a parameter change ρ such that $\tilde{f} = f \circ \rho$ and $\tilde{g} = g \circ \rho$ in some neighbourhood of 0. By the inverse function theorem of complex analysis (see Appendix B) this is an equivalence relation on the set of pairs. The equivalence class of a pair (f, g) is called a meromorphic element; *we shall use the notation $< f, g >$ or $< f(t), g(t) >$ for it.*

We let $\mathcal{M}$ denote the set of all meromorphic elements. We shall make $\mathcal{M}$ into a Riemann surface with infinitely many connected components. Each connected component will be the Riemann surface associated to some holomorphic function. First we must define a topology on $\mathcal{M}$.

Let (f, g) be a pair and let $r > 0$ be sufficiently small that f and g are both defined and meromorphic on the open disc $D(0, r)$ of centre 0 and radius r, and the map

$$t \mapsto (f(t), g(t))$$

is one-to-one on $D(0, r)$). Then $(f(t_0 + t), g(t_0 + t))$ is a pair for each $t_0 \in D(0, r)$ and so we can define

$$U(f, g, r) = \{< f(t_0 + t), g(t_0 + t) > : t_0 \in D(0, r)\} \subseteq \mathcal{M}.$$

Note that $< f, g > \in U(f, g, r)$.

Lemma 7.2 *There is a topology on $\mathcal{M}$ such that a subset of $\mathcal{M}$ is open if and only if it is a union of subsets of the form $U(f, g, r)$ just defined.*

Proof. It suffices to show that if

$$< f, g > \in U(f_1, g_1, r_1) \cap U(f_2, g_2, r_2)$$

then there is some $r > 0$ such that

$$U(f, g, r) \subseteq U(f_1, g_1, r_1) \cap U(f_2, g_2, r_2).$$

So suppose that

$$< f, g > = < f_1(t_1 + t), g_1(t_1 + t) > = < f_2(t_2 + t), g_2(t_2 + t) >$$

for some $t_1 \in D(0, r_1)$ and $t_2 \in D(0, r_2)$. Then there are parameter changes ρ_1 and ρ_2 such that

$$f_1(t_1 + \rho_1(t)) = f(t) = f_2(t_2 + \rho_2(t))$$

and

$$g_1(t_1 + \rho_1(t)) = g(t) = g_2(t_2 + \rho_2(t))$$

for all $t \in D(0, s)$ where $s \leq min(r_1, r_2)$. Since ρ_1 and ρ_2 are parameter changes there exists $r > 0$ such that $r \leq s$ and on the open disc $D(0, r)$ both ρ_1 and ρ_2 are holomorphic and ρ_1' and ρ_2' do not vanish, and also

$$\rho_1(D(0, r)) \subseteq D(0, r_1 - |t_1|)$$

and

$$\rho_2(D(0, r)) \subseteq D(0, r_2 - |t_2|).$$

Then if $t_0 \in D(0, r)$, the function

$$\sigma(t) = \rho_1(t_0 + t) - \rho_1(t_0)$$

is a parameter change and so

$$
\begin{aligned}
< f(t_0 + t), g(t_0 + t) > &= \ < f_1(t_1 + \rho_1(t_0 + t)), g_1(t_1 + \rho_1(t_0 + t)) > \\
&= \ < f_1(t_1 + \rho_1(t_0) + \sigma(t)), g_1(t_1 + \rho_1(t_0) + \sigma(t)) > \\
&= \ < f_1(t_1 + \rho_1(t_0) + t), g_1(t_1 + \rho_1(t_0) + t) >
\end{aligned}
$$

which lies in $U(f_1, g_1, r_1)$ since $| t_1 + \rho_1(t_0) | < r_1$. Thus

$$U(f, g, r) \subseteq U(f_1, g_1, r_1)$$

and similarly

$$U(f, g, r) \subseteq U(f_2, g_2, r_2)$$

as required. $\square$

In order to show that $\mathcal{M}$ is a Riemann surface we need to find an atlas $\{\phi_\alpha : U_\alpha \to V_\alpha \ : \ \alpha \in \mathcal{A}\}$ of holomorphic charts on $\mathcal{M}$.

Definition 7.3 *Let $\mathcal{A}$ be the set of all ordered triples (f, g, r) where (f, g) is a pair and $r > 0$ is sufficiently small that f and g are defined and meromorphic on $D(0, r)$ and the mapping*

$$t \mapsto (f(t), g(t))$$

on $D(0, r)$ is one-to-one.

Lemma 7.4 *If $(f, g, r) \in \mathcal{A}$ then the map from $D(0, r)$ to $U(f, g, r)$ defined by*

$$t_0 \mapsto < f(t_0 + t), g(t_0 + t) >$$

is a homeomorphism.

Proof. The map is clearly surjective by the definition of $U(f, g, r)$. It is injective since if

$$< f(t_0 + t), g(t_0 + t) > = < f(t_1 + t), g(t_1 + t) >$$

then $f(t_0) = f(t_1)$ and $g(t_0) = g(t_1)$ so t_0 and t_1 are mapped to the same point by the injective map

$$t \mapsto (f(t), g(t)).$$

That it is a homoeomorphism follows straight from the definition of the topology on $\mathcal{M}$, since the image of any open disc $D(a, \varepsilon)$ contained in $D(0, r)$ is the subset $U(f(a + t), g(a + t), \varepsilon)$ of $\mathcal{M}$. □

Definition 7.5 *If $\alpha = (f, g, r) \in \mathcal{A}$ let*

$$U_\alpha = U(f, g, r)$$

and

$$V_\alpha = D(0, r)$$

and let $\phi_\alpha : U_\alpha \to V_\alpha$ be the inverse of the homeomorphism of lemma 7.4.

Proposition 7.6 *$\mathcal{M}$ is a Riemann surface with the holomorphic atlas*

$$\Phi = \{\phi_\alpha : U_\alpha \to V_\alpha \ : \ \alpha \in \mathcal{A}\}$$

defined at 7.3 and 7.5.

Proof. We shall leave the proof that $\mathcal{M}$ is Hausdorff to the end of the section (see lemma 7.24). Thus we just need to show that if $\alpha = (f_1, g_1, r_1)$ and $\beta = (f_2, g_2, r_2)$ belong to the indexing set $\mathcal{A}$ then the transition function $\phi_\beta \circ \phi_\alpha^{-1}$ is holomorphic near every point of $\phi_\alpha(U_\alpha \cap U_\beta)$.

Let u be any point of $U_\alpha \cap U_\beta$, i.e.

$$u =< f_1(t_1 + t), g_1(t_1 + t) >=< f_2(t_2 + t), g_2(t_2 + t) >$$

where $t_1 = \phi_\alpha(u)$ and $t_2 = \phi_\beta(u)$. There is a parameter change ρ such that if t is sufficiently near 0 then

$$f_1(t_1 + t) = f_2(t_2 + \rho(t))$$

and

$$g_1(t_1 + t) = g_2(t_2 + \rho(t)).$$

If t_0 is close to $t_1 = \phi_\alpha(u)$ then $\phi_\alpha^{-1}(t_0) =< f_1(t_0 + t), g_1(t_0 + t) >$. In order to work out $\phi_\beta \phi_\alpha^{-1}$ we write

$$
\begin{aligned}
f_1(t_0 + t) &= f_1(t_1 + (t_0 - t_1 + t)) \\
&= f_2(t_2 + \rho(t_0 - t_1 + t)) \\
&= f_2(t_2 + \rho(t_0 - t_1) + \sigma(t))
\end{aligned}
$$

where

$$\sigma(t) = \rho(t_0 - t_1 + t) - \rho(t_0 - t_1)$$

satisfies

$$\sigma(0) = 0$$

and

$$\sigma'(0) = \rho'(t_0 - t_1).$$

Similarly

$$g_1(t_0 + t) = g_2(t_2 + \rho(t_0 - t_1) + \sigma(t)).$$

Since $\rho'(0) \neq 0$ it follows that if $t_0 - t_1$ is small enough then $\sigma'(0) \neq 0$ so that σ is a parameter change. This means that

$$
\begin{aligned}
\phi_\alpha^{-1}(t_0) &= < f_1(t_0 + t), g_1(t_0 + t) > \\
&= < f_2(t_2 + \rho(t_0 - t_1) + \sigma(t)), g_2(t_2 + \rho(t_0 - t_1) + \sigma(t)) > \\
&= < f_2(t_2 + \rho(t_0 - t_1) + t, g_2(t_2 + \rho(t_0 - t_1) + t >
\end{aligned}
$$

so that

$$\phi_\beta \phi_\alpha^{-1}(t_0) = t_2 + \rho(t_0 - t_1).$$

This is true for all t_0 sufficiently close to t_1, where t_1 and t_2 are fixed. Since ρ is holomorphic near 0 it follows that the transition function $\phi_\beta \phi_\alpha^{-1}$ is holomorphic near $t_1 = \phi_\alpha(u)$, as required. $\square$

Remark 7.7 It is easy to see that the functions $\psi : \mathcal{M} \to \mathbf{C} \cup \{\infty\}$ and $\chi : \mathcal{M} \to \mathbf{C} \cup \{\infty\}$ defined by

$$\psi(< f, g >) = f(0)$$

and

$$\chi(< f, g >) = g(0)$$

are meromorphic functions on $\mathcal{M}$. To show this it suffices to show that their compositions with the inverses of the holomorphic charts $\phi_\alpha : U_\alpha \to V_\alpha$ are meromorphic for each $\alpha = (f, g, r) \in \mathcal{A}$. This is true since f and g are meromorphic functions on $V_\alpha = D(0, r)$ and

$$\psi \circ \phi_\alpha^{-1}(t_0) = \psi(< f(t_0 + t), g(t_0 + t) >) = f(t_0)$$

and

$$\chi \circ \phi_\alpha^{-1}(t_0) = \chi(< f(t_0 + t), g(t_0 + t) >) = g(t_0)$$

for all $t_0 \in D(0, r)$.

$\mathcal{M}$ is the disjoint union of its connected components, each of which is an open subset of $\mathcal{M}$ and therefore is itself a Riemann surface by 5.40 (b).

Definition 7.8 *If f is a meromorphic function defined on a neighbourhood of a point $a \in \mathbf{C} \cup \{\infty\}$ then the Riemann surface of f is the connected component of $\mathcal{M}$ containing the meromorphic element*

$$< a + t, f(a + t) >$$

if $a \in \mathbf{C}$ or

$$< t^{-1}, f(t^{-1}) >$$

if $a = \infty$. This meromorphic element is called the germ of f at a and is written $[f]_a$.

Remark 7.9 The concept of the Riemann surface of a holomorphic function is closely related to the concept of analytic continuation of a holomorphic function along a path (see e.g. [Griffiths 89]II§2).

Remark 7.10 Let (f, g) be a pair. If f is holomorphic at 0 then f has a Taylor expansion of the form

$$f(t) = \sum_{n \geq 0} c_n t^n$$

near 0. Since f is not constant we can define m to be the multiplicity of the zero of $f(t) - c_0$ at 0, i.e. the smallest positive integer such that $c_m \neq 0$. Then

$$f(t) = c_0 + t^m h(t)$$

where

$$h(t) = c_m + c_{m+1}t + c_{m+2}t^2 + \cdots$$

is holomorphic near 0. Since $h(0) = c_m \neq 0$ we can find a holomorphic mth root $k(t)$ of $h(t)$ defined near 0 and then

$$\rho(t) = t\,k(t)$$

is a parameter change such that

$$f(t) = c_0 + \rho(t)^m.$$

Thus

$$< f, g > = < c_0 + t^m, g \circ \rho^{-1}(t) > .$$

Similarly if f has a pole of order m at 0 then f has a Laurent expansion

$$f(t) = \sum_{n \geq -m} c_n t^n = t^{-m} h(t)$$

near 0 where $c_{-m} \neq 0$ and

$$h(t) = c_{-m} + c_{-m+1}t + c_{-m+2}t^2 + \cdots$$

is holomorphic near 0. We can find a holomorphic mth root of $1/h(t)$ and multiply by t to get a parameter change $\rho(t)$ such that

$$f(t) = \rho(t)^{-m},$$

so that

$$< f, g > = < t^{-m}, g \circ \rho^{-1}(t) > .$$

This means that every element $< f, g >$ of $\mathcal{M}$ can be expressed in the form

$$< a + t^m, g(t) >$$

or

$$< t^{-m}, g(t) >$$

for some positive integer m and meromorphic function g near 0. If $m = 1$ we have the germ of g at a (or at ∞). If $m > 1$ then the meromorphic element $< f, g >$ is called a *branch point* of the component of $\mathcal{M}$ to which it belongs (this is the traditional terminology, though "ramification point" would fit more precisely with the usage in §4.2).

Now we can relate projective curves in $\mathbf{P_2}$ to components of $\mathcal{M}$.

Definition 7.11 *Let $P(x,y,z)$ be a nonconstant irreducible homogeneous polynomial of degree d not divisible by z. The Riemann surface S_P of $P(x,y,z)$ is the open subset of $\mathcal{M}$ consisting of all those elements $<f,g>$ of $\mathcal{M}$ satisfying*

$$P(f(t), g(t), 1) = 0$$

for all t in some neighbourhood of 0. If C is the projective curve

$$C = \{[x,y,z] \in \mathbf{P}_2 \; : \; P(x,y,z) = 0\}$$

then we write $\tilde{C}$ for S_P and define $\pi : \tilde{C} \to C$ by

$$\pi(<f,g>) = [f(0), g(0), 1]$$

if f and g are both holomorphic near 0, and otherwise

$$\pi(<f,g>) = [\tilde{f}(0), \tilde{g}(0), 0]$$

where $\tilde{f}(t) = t^n f(t)$ and $\tilde{g}(t) = t^n g(t)$ and n is the multiplicity of the pole at 0 of f or g, whichever is the greater.

Our aim is to prove the following theorem.

Theorem 7.12 *$\tilde{C}$ is a compact connected Riemann surface. The map $\pi : \tilde{C} \to C$ is continuous and surjective. If C is nonsingular then π is a holomorphic bijection, and in general $\pi^{-1}(Sing(C))$ is finite and*

$$\pi : \tilde{C} - \pi^{-1}(Sing(C)) \to C - Sing(C)$$

is a holomorphic bijection.

Remark 7.13 In fact the map $\pi : \tilde{C} \to C$ is holomorphic even at points of $\pi^{-1}(Sing(C))$, but we have not discussed how to make sense of this.

Remark 7.14 The only place in the proof of this theorem where the irreducibility of $P(x,y,z)$ will be needed is the connectedness of $\tilde{C}$. In fact $\tilde{C}$ is connected if and only if $P(x,y,z)$ is irreducible.

As a first step towards the proof of theorem 7.12 let us check that $\pi : \tilde{C} \to C$ is continuous.

Lemma 7.15 *$\pi : \tilde{C} \to C$ is continuous and its restriction to $\tilde{C} - \pi^{-1}Sing(C)$ is holomorphic.*

Proof. π is continuous at any $< f_0, g_0 > \in \tilde{C}$ such that f_0 and g_0 are both holomorphic near 0, because locally it is the composition of the continuous map $\pi : \mathbf{C}^3 - \{0\} \to \mathbf{P}_2$ defined by $\pi(x, y, z) = [x, y, z]$ with the map $\tilde{C} \to \mathbf{C}^3 - \{0\}$ defined by

$$< f, g > \mapsto (f(0), g(0), 1) = (\psi(< f, g >), \chi(< f, g >), 1)$$

where ψ and χ are holomorphic near $< f, g >$ by remark 7.7. If f_0 has a pole at 0 of multiplicity greater than or equal to that of g_0 then near $< f_0, g_0 >$ the map π is the composition of $\pi : \mathbf{C}^3 - \{0\} \to \mathbf{P}_2$ with the map

$$< f, g > \mapsto (1, \frac{g(0)}{f(0)}, \frac{1}{f(0)})$$

which is continuous since χ/ψ and $1/\psi$ are holomorphic near $< f_0, g_0 >$, so π is continuous at $< f_0, g_0 >$. Similarly if g_0 has a pole at 0 of multiplicity greater than that of f_0 then π is continuous at $< f_0, g_0 >$.

To show that the restriction of π to $\tilde{C} - \pi^{-1}(Sing(C))$ is holomorphic consider any $\alpha = (f, g, r) \in \mathcal{A}$. Then the holomorphic chart $\phi_\alpha : U_\alpha \to V_\alpha$ has inverse

$$t_0 \mapsto < f(t_0 + t), g(t_0 + t) >$$

defined on $V_\alpha = D(0, r)$. A holomorphic chart ψ_β on $C - Sing(C)$ is given by one of the functions taking $[x, y, z]$ to $\frac{x}{z}, \frac{y}{z}, \frac{z}{y}, \frac{x}{y}, \frac{y}{x}$ or $\frac{z}{x}$. The composition $\psi_\beta \circ \pi \circ \phi_\alpha^{-1}$ is therefore given where it is defined by one of the functions $f, g, \frac{f}{g}, \frac{1}{g}, \frac{g}{f}$ or $\frac{1}{f}$. These functions are all meromorphic, that is, holomorphic as functions into $\mathbf{C} \cup \{\infty\}$. Since holomorphic charts always take values in $\mathbf{C}$ it follows that where it is defined the composition $\psi_\beta \circ \pi \circ \phi_\alpha^{-1}$ is always a holomorphic function from an open subset of $\mathbf{C}$ into $\mathbf{C}$. Therefore the restriction of π to $\tilde{C} - \pi^{-1}(Sing(C))$ is holomorphic. $\square$

Lemma 7.16 *The restriction of $\pi : \tilde{C} \to C$ to $\tilde{C} - \pi^{-1}(Sing(C))$ is a bijection onto $C - Sing(C)$.*

Proof. Suppose that $[a, b, c]$ is a nonsingular point of C with $c \neq 0$. We can assume $c = 1$. Then either $\frac{\partial P}{\partial x}$ or $\frac{\partial P}{\partial y}$ is nonzero at $(a, b, 1)$, for if both were zero then by Euler's relation 2.32 $\frac{\partial P}{\partial z}$ would also vanish and so $[a, b, 1]$ would be a singular point. We can suppose without loss of generality that

$$\frac{\partial P}{\partial y}(a, b, 1) \neq 0;$$

we wish to show that there is a unique $< f, g > \in \tilde{C}$ which π maps to $[a, b, 1]$. This comes from the implicit function theorem (see Appendix B) applied to the polynomial $P(x, y, 1)$ which tells us that there are open neighbourhoods

U and V of a and b in $\mathbf{C}$ and a holomorphic function $h : U \to V$ such that if $x \in U$ and $y \in V$ then

$$P(x, y, 1) = 0$$

if and only if $y = h(x)$. This means that if f and g are holomorphic near 0 and satisfy $f(0) = a$ and $g(0) = b$ and

$$P(f(t), g(t), 1) = 0$$

for all t near 0 we must have

$$g(t) = h(f(t))$$

for all t near 0. Then if (f, g) is a pair it follows that f must be injective and so

$$\rho(t) = f(t) - a$$

is a parameter change. Hence

$$< f, g >=< a + \rho(t), h(a + \rho(t)) >=< a + t, h(a + t) >$$

and this is the unique element of $\tilde{C}$ satisfying $\pi(< f, g >) = [a, b, 1]$.

The argument if $c = 0$ is similar; we apply the implicit function theorem to the polynomial $P(x, 1, z)$ or $P(1, y, z)$ instead of $P(x, y, 1)$, and consider the functions $\frac{f}{g}$ and $\frac{1}{g}$ or $\frac{g}{f}$ and $\frac{1}{f}$ instead of f and g. $\square$

Lemma 7.17 $\pi : \tilde{C} \to C$ *is surjective.*

Proof. Fix $[a, b, c] \in C$. We need to show that $\pi^{-1}([a, b, c])$ is nonempty. We can assume by an appropriate choice of coordinates that $[0, 1, 0] \notin C$. Then in particular $\frac{\partial P}{\partial y}$ is not identically zero, so by Bézout's theorem 3.3 as C is irreducible there are at most finitely many points of C where $\frac{\partial P}{\partial y}$ is zero.

Suppose first that $c \neq 0$; then we can take $c = 1$. Then there exists $\varepsilon > 0$ such that if $x \in C$ and $0 <| a - x |\leq \varepsilon$ then there is no $y \in \mathbf{C}$ such that $[x, y, 1] \in C$ and

$$\frac{\partial P}{\partial y}(x, y, 1) = 0.$$

Let $D_\pm(a, \varepsilon)$ be the open disc $D(a, \varepsilon)$ in $\mathbf{C}$ with the straight line segment from a to $a \pm \varepsilon$ removed. Since $D_\pm(a, \varepsilon)$ is simply connected, it follows from C.6 of Appendix C that the holomorphic map $\phi : C \to \mathbf{P}_1 = \mathbf{C} \cup \{\infty\}$ defined by $\phi([x, y, z]) = [x, z]$, or equivalently $\phi([x, y, 1]) = x$, restricts to a homeomorphism from each connected component of

$$\phi^{-1}(D_\pm(a, \varepsilon))$$

onto $D_\pm(a,\varepsilon)$ and there are d connected components. By the inverse function theorem (see Appendix B) the inverse of the restriction of ϕ to the jth component of $\phi^{-1}(D_\pm(a,\varepsilon))$ is holomorphic. It must be of the form

$$x \mapsto [x, f_j^\pm(x), 1]$$

where

$$P(x, f_j^\pm(x), 1) = 0$$

for all $x \in D_\pm(a,\varepsilon)$. Moreover $f_j^\pm(x) \neq f_i^\pm(x)$ if $i \neq j$ since $[x, f_j^\pm(x), 1]$ and $[x, f_i^\pm(x), 1]$ lie in different connected components of $\phi^{-1}(D_\pm(a,\varepsilon))$. Also since C is compact each $f_j^\pm$ is bounded on $D_\pm(a,\varepsilon)$ and its only possible limiting values as $x \to a$ are the finitely many values of $y \in C$ such that $[a, y, 1] \in C$. As b is one of these values so that $[a, b, 1]$ belongs to the closure of $\phi^{-1}(D_\pm(a,\varepsilon))$ we can assume that b is a limiting value of $f_1^\pm$ as $x \to a$.

Since $D_+(a,\varepsilon) \cap D_-(a,\varepsilon)$ has two connected components, the lower and upper halves of the disc $D(a,\varepsilon)$, we can reorder $f_1^+, \ldots, f_d^+$ and $f_1^-, \ldots, f_d^-$ so that if $1 \leq j \leq d$ then f_j^+ agrees with f_j^- on the lower half of the disc and with $f_{\sigma(j)}^-$ on the upper half of the disc, where σ is a permutation of $\{1, \ldots, d\}$. Moreover σ can be assumed to satisfy

$$\sigma(i) = i + 1$$

if $1 \leq i < m$ and

$$\sigma(m) = 1$$

for some $m \leq d$. Therefore there is a well-defined holomorphic function

$$g : \{t \in \mathbf{C} : 0 < |t| < \varepsilon^{\frac{1}{m}}\} \to \mathbf{C}$$

given by

$$g(t) = f_j^+(a + t^m)$$

if $(2j - 2)\pi/m < arg(t) < 2j\pi/m$ and

$$g(t) = f_j^-(a + t^m)$$

if $(2j - 1)\pi/m < arg(t) < (2j + 1)\pi/m$. Since g is bounded and has b as a limiting value as $t \to 0$ it follows that g can be extended to a holomorphic function on the disc $D(0, \varepsilon^{\frac{1}{m}})$ such that $g(0) = b$ (see e.g. [Priestley 85] 6.12). Moreover

$$P(a + t^m, g(t), 1) = 0$$

for all $t \in D(0, \varepsilon^{\frac{1}{m}})$. Finally the mapping

$$t \mapsto (a + t^m, g(t))$$

is injective on $D(0, \varepsilon^{\frac{1}{m}})$ since $f_j^{\pm}(x) \neq f_i^{\pm}(x)$ if $i \neq j$. Thus $< a + t^m, g(t) > \in \tilde{C}$ and $\pi(< a + t^m, g(t) >) = [a, b, 1]$.

Now suppose $c = 0$; then $a \neq 0$ since $[0, 1, 0] \notin C$ so we may assume $a = 1$. There exists $\varepsilon > 0$ such that if $x \in \mathbf{C}$ and $|x| > \frac{1}{\varepsilon}$ then there is no $y \in \mathbf{C}$ such that $[x, y, 1] \in C$ and

$$\frac{\partial P}{\partial y}(x, y, 1) = 0.$$

Let

$$D_{\pm}(\infty, \varepsilon) = \{x \in \mathbf{C} : |x| > \frac{1}{\varepsilon}, \ x \notin \mathbf{R}_{\pm}\}$$

where $\mathbf{R}_+$ and $\mathbf{R}_-$ are the positive and negative real axes in $\mathbf{C}$. Now we can repeat the previous argument, replacing a by ∞ and $a + t^m$ by t^{-m} to show that there exists an element of $\tilde{C}$ of the form $< t^{-m}, g(t) >$ which is mapped by π to $[a, b, c] = [1, b, 0]$. $\square$

Remark 7.18 Recall from 7.7 that the composition $\psi = \phi \circ \pi : \tilde{C} \to \mathbf{P}_1$ given by $< f, g > \mapsto f(0)$ is a meromorphic function on $\tilde{C}$. Decompose the permutation σ obtained in the proof of lemma 7.17 into disjoint cycles

$$\sigma = \sigma_1 \ldots \sigma_l$$

where σ_i is a cycle of length $m_j \geq 1$ and $m_1 + \cdots + m_l = d$. The proof of lemma 7.17 then shows that if $a \in \mathbf{C}$ there are holomorphic functions $g_1, \ldots, g_l$ near 0 and pairs

$$< a + t^{m_j}, g_j(t) > \in \tilde{C}$$

such that

$$\psi^{-1}(a) = \{< a + t^{m_j}, g_j(t) > : 1 \leq j \leq l\}.$$

Moreover ψ takes the value a at $< a + t^{m_j}, g_j(t) >$ with multiplicity m_j in the sense that the multiplicity of the zero of $\psi - a$ at $< a + t^{m_j}, g_j(t) >$ is m_j. Thus for any $a \in \mathbf{C}$ the meromorphic function ψ on $\tilde{C}$ takes the value a exactly d times, counting multiplicities. A similar argument shows that this is also true when $a = \infty$.

Remark 7.19 The last remark tells us that there are holomorphic functions $g_1, \ldots, g_l$ defined near 0 and positive integers $m_1, \ldots, m_l$ such that $m_1 + \cdots + m_l = d$ and

$$P(a + t^{m_j}, g_j(t), 1) = 0$$

for all t near 0. In other words, if $(x - a)^{\frac{1}{m_j}}$ is any m_jth root of $(x - a)$ then when x is close enough to a

$$y = g_j((x - a)^{\frac{1}{m_j}})$$

is a solution to the equation

$$P(x, y, 1) = 0$$

regarded as defining y as a multivalued function of x. Furthermore if $|x-a|$ is sufficiently small but nonzero then by the proof of lemma 7.17 the d numbers

$$y_j = g_j(e^{2\pi i s/m_j}(x-a)^{1/m_j})$$

for $1 \leq i \leq l$ and $1 \leq s \leq m_j$ are distinct (since the points $[x, y_j, 1]$ lie in different connected components of $\phi^{-1}(D_{\pm}(a, \varepsilon))$). Since $P(x, y, z)$ is a polynomial of degree d it follows that for any y and any x sufficiently close to a we can write

$$P(x, y, 1) = K \prod_{1 \leq j \leq l} \prod_{1 \leq s \leq m_j} (y - g_j(e^{2\pi i s/m_j}(x-a)^{1/m_j}))$$

for some constant K.

Now we can prove that $\tilde{C}$ is compact.

Lemma 7.20 $\tilde{C}$ *is compact.*

Proof. By remark 7.18 it suffices to prove the following result.

Lemma 7.21 *Let* $\Phi : S \to \mathbf{P}_1$ *be a meromorphic function on a Riemann surface* S *which takes any value* $a \in \mathbf{P}_1 = \mathbf{C} \cup \{\infty\}$ *exactly d times, counting multiplicities. Then S is compact.*

Proof. Since $\mathbf{P}_1$ is compact by 2.18 it suffices to show that every $a \in \mathbf{P}_1 = \mathbf{C} \cup \{\infty\}$ has an open neighbourhood W_a in $\mathbf{P}_1$ such that $\Phi^{-1}(W_a)$ is contained in a compact subset S_a of S. For then the open cover $\{W_a : a \in \mathbf{P}_1\}$ of $\mathbf{P}_1$ has a finite subcover, say

$$\mathbf{P}_1 = W_{a_1} \cup \ldots \cup W_{a_k},$$

and so

$$\begin{aligned} S &= \Phi^{-1}(W_{a_1}) \cup \ldots \cup \Phi^{-1}(W_{a_k}) \\ &= S_{a_1} \cup \ldots \cup S_{a_k} \end{aligned}$$

which is a finite union of compact subsets and hence is compact.

For any $a \in \mathbf{C} \cup \{\infty\}$ let $\Phi^{-1}(a) = \{s_1, \ldots, s_l\}$ where Φ takes the value a at s_j with multiplicity m_j and $m_1 + \cdots + m_l = d$. For simplicity let us assume $a \neq \infty$; if $a = \infty$ we can apply the same argument to $1/\Phi$. Let $\phi_j : U_j \to V_j$

be a holomorphic chart on S near s_j such that $\phi_j(s_j) = 0$ and such that $U_1, \ldots, U_l$ are disjoint. Then the holomorphic function f_j defined near 0 by

$$f_j(z) = \Phi(\phi_j^{-1}(z)) - a$$

has a zero of multiplicity m_j at 0. Choose $\varepsilon_j > 0$ such that the closed disc

$$\overline{D(0,\varepsilon_j)} = \{z \in \mathbf{C} : |z| \le \varepsilon\}$$

is contained in V_j. Then by remark B.2 and theorem B.3 of Appendix B there exists $\delta_j > 0$ such that if $|a' - a| < \delta_j$ then the function

$$f_j - a' + a = \Phi \circ \phi_j^{-1} - a'$$

has at least m_j zeros counting multiplicities in the open disc $D(0,\varepsilon_j)$. Let

$$W_a = \{z \in \mathbf{C} : |z - a| < min\{\delta_1, \ldots, \delta_l\}\}$$

and let

$$S_a = \bigcup_{1 \le j \le l} \phi_j^{-1}(\overline{D(0,\varepsilon_j)}).$$

Then S_a is compact because ϕ_j is a homeomorphism and $\overline{D(0,\varepsilon_j)}$ is compact. Moreover if $a' \in W_a$ then Φ takes the value a' at least m_j times counting multiplicities in the open subset $\phi_j^{-1}(D(0,\varepsilon_j))$ of U_j. Since $U_1, \ldots, U_l$ are disjoint and $m_1 + \cdots + m_l = d$ it follows that Φ takes the value a' at least d times counting multiplicities in S_a, and hence

$$\Phi^{-1}(a') \subseteq S_a.$$

This is true for all $a' \in W_a$ so

$$\Phi^{-1}(W_a) \subseteq S_a$$

as required. $\square$

Proposition 7.22 $\tilde{C}$ *and* C *are connected.*

Proof. By lemma 7.17 it suffices to show that $\tilde{C}$ is connected, since the image of a connected space under a continuous map is always connected.

Let A be a connected component of $\tilde{C}$; we wish to show that $\tilde{C} = A$. As in the proof of lemma 7.17 we choose coordinates so that $[0,1,0] \notin C$ and consider the map $\phi : C \to \mathbf{P}_1$ defined by

$$\phi([x,y,z]) = [x,z].$$

If $a \in \mathbf{P}_1 = \mathbf{C} \cup \{\infty\}$ let $D_\pm(a, \varepsilon)$ be defined as in the proof of lemma 7.17. Then $\phi^{-1}(D_\pm(a, \varepsilon))$ has d connected components and the restriction of ϕ to the jth component is a homeomorphism onto $D_\pm(a, \varepsilon)$ with inverse

$$x \mapsto [x, f_j^\pm(x), 1]$$

where $f_j^\pm$ is holomorphic on $D_\pm(a, \varepsilon)$ for $1 \leq j \leq d$. We can assume that f_j^+ agrees with f_j^- on the lower half of the disc $D(a, \varepsilon)$ and agrees with $f_{\sigma(j)}^-$ on the upper half, where σ is a permutation of $\{1, \ldots, d\}$. Each connected component of $\phi^{-1}(D_\pm(a, \varepsilon))$ is either contained in or disjoint from A, so we can assume that for some $e \in \{1, \ldots, d\}$

$$f_j^\pm(x) \in A$$

if and only if $j \in \{1, \ldots, e\}$, and σ restricts to a permutation of $\{1, \ldots, e\}$. In order to prove that $\tilde{C} = A$ it suffices to show that $e = d$ (for every choice of $a \in \mathbf{C} \cup \{\infty\}$).

As in remarks 7.18 and 7.19 we can decompose σ as a product of disjoint cycles

$$\sigma_1 \ldots \sigma_l$$

of lengths $m_1, \ldots, m_l$ such that $m_1 + \cdots + m_l = d$ and if $a \neq \infty$ we can find holomorphic functions $g_1, \ldots, g_l$ near 0 and a nonzero constant K such that

$$P(x, y, 1) = K \prod_{1 \leq j \leq l} \prod_{1 \leq s \leq m_j} (y - g_j(e^{2\pi i s/m_j}(x - a)^{1/m_j}))$$

for all y and all x near a. (If $a = \infty$ we can simply reverse the roles of x and z). We may assume that the restriction of σ to $\{1, \ldots, e\}$ is given by the product

$$\sigma_1 \ldots \sigma_k$$

for some $k \leq l$ such that $m_1 + \cdots + m_l = e$. Now consider the function

$$Q(x, y, 1) = K \prod_{1 \leq j \leq k} \prod_{1 \leq s \leq m_j} (y - g_j(e^{2\pi i s/m_j}(x - a)^{1/m_j})).$$

This is a polynomial of degree e in y whose coefficients are power series in the expressions

$$e^{2\pi i s/m_j}(x - a)^{1/m_j}$$

which converge for x near a. Moreover the power series are unchanged by permutations of these expresssions and hence can be expressed in terms of the elementary symmetric polynomials which are (up to signs) the coefficients of the polynomial

$$\prod_{1 \leq j \leq k} \prod_{1 \leq s \leq m_j} (y - e^{2\pi i s/m_j}(x - a)^{1/m_j}) = \prod_{1 \leq j \leq k} (y^{m_j} - x + a)$$

in y. These coefficients are constant multiples of powers of $x - a$, so $Q(x,y)$ is a polynomial in y whose coefficients are power series in $x - a$ which converge to holomorphic functions for x close to a.

A priori the definition of $Q(x,y)$ depends on the choice of $a \in \mathbb{C}$. However if $x \in D_\pm(a,\varepsilon)$ then

$$A \cap \phi^{-1}\{x\} = \{[x, g_j(e^{2\pi i s/m_j}(x-a)^{1/m_j}), 1] : 1 \le j \le k,\ 1 \le s \le m_j\}.$$

From this it follows that $Q(x,y)$ is unchanged if we replace a by $a' \in D(a,\varepsilon)$, and hence, since $\mathbb{C}$ is connected, that $Q(x,y)$ is independent of the choice of $a \in \mathbb{C}$. In particular $Q(x,y)$ is a polynomial in y whose coefficients are holomorphic functions of $x \in \mathbb{C}$.

Next by considering $a = \infty$ we find that the coefficients of powers of y in $Q(x,y)$ extend to meromorphic functions on $\mathbf{P}_1 = \mathbb{C} \cup \{\infty\}$ whose only poles are at ∞. We have shown earlier (5.41) that the only meromorphic functions on $\mathbf{P}_1$ are rational functions, and hence a meromorphic function on $\mathbf{P}_1$ whose only poles are at ∞ must be a polynomial function. Thus $Q(x,y)$ is a polynomial in x and y.

The same argument shows that

$$\prod_{k<j\le l} \prod_{1\le s\le m_j} (y - g_j(e^{2\pi i s/m_j}(x-a)^{1/m_j}))$$

is also a polynomial in x and y, and hence that $Q(x,y)$ divides $P(x,y,1)$. It follows that the associated homogeneous polynomial

$$z^e Q(\frac{x}{z}, \frac{y}{z})$$

divides $P(x,y,z)$. But $Q(x,y)$ is nonconstant and $P(x,y,z)$ is irreducible, so $Q(x,y)$ must be a constant multiple of $P(x,y,1)$. Thus $k = l$ and $e = d$ so $A = \tilde{C}$ as required. $\square$

Remark 7.23 In fact every compact component S of $\mathcal{M}$ is of the form S_P for some $P(x,y,z)$, and hence is the resolution of singularities $\tilde{C}$ of a projective curve C in $\mathbf{P}_2$. The proof runs as follows. First the compactness of S is used to show that the meromorphic function $\psi : S \to \mathbf{P}_1$ defined as at 7.7 by

$$\psi(< f, g >) = f(0)$$

takes every value the same number of times, counting multiplicities. Let this number be d. Then one shows that the subset F of $\mathbb{C}$ consisting of those values b taken by ψ with multiplicity at least two at some point of S is finite, and that if $\chi : S \to \mathbf{P}_1$ is defined as at 7.7 then the function

$$Q(z,w) = \prod_{s\in S, \psi(s)=z} (w - \chi(s)),$$

defined for $z \in \mathbf{C} - F$ and $w \in \mathbf{C}$, extends to a polynomial function on $\mathbf{C} \times \mathbf{C}$ of the form $P(z, w, 1)$ where $S = S_P$.

This leads to a proof that *any* compact Riemann surface S is the resolution of singularities $\tilde{C}$ of an irreducible projective curve C in $\mathbf{P}_2$. The crucial ingredient which it is beyond the scope of this book to prove is that if q and r are distinct points of S then there is a meromorphic function f on S such that $f(q) \neq f(r)$. Using this it is possible (cf. exercise 7.3) to construct a pair of meromorphic functions f and g on S such that there is an isomorphism from S onto a connected component of $\mathcal{M}$ given by

$$p \mapsto < f \circ \phi_p^{-1}, g \circ \phi_p^{-1} >$$

for each $p \in S$, where ϕ_p is a holomorphic chart on S such that $\phi_p(p) = 0$. For more details see e.g. [Jones 71] chapters V and VI.

It remains for us to prove

Lemma 7.24 $\mathcal{M}$ *is Hausdorff.*

Proof. Let us suppose that $< f, g >$ and $< \tilde{f}, \tilde{g} >$ are elements of $\mathcal{M}$ which have no distinct neighbourhoods in $\mathcal{M}$. We need to show that $< f, g >$ is equal to $< \tilde{f}, \tilde{g} >$.

By remark 7.10 we can assume that $f(t)$ is $a + t^m$ or t^{-m} and $\tilde{f}(t)$ is $b + t^n$ or t^{-n} for some $a, b \in \mathbf{C}$ and positive integers m, n.

If k is a large enough positive integer then $U(f, g, \frac{1}{k})$ and $U(\tilde{f}, \tilde{g}, \frac{1}{k})$ are neighbourhoods of $< f, g >$ and $< \tilde{f}, \tilde{g} >$ in $\mathcal{M}$ and hence by assumption have a common element, say

$$< f(s_k + t), g(s_k + t) > = < \tilde{f}(t_k + t), \tilde{g}(t_k + t) >$$

for some $s_k, t_k \in D(0, \frac{1}{k})$. In particular

$$f(s_k) = \tilde{f}(t_k)$$

and

$$g(s_k) = \tilde{g}(t_k).$$

Now letting $k \to \infty$ shows that $f(0) = \tilde{f}(0)$, so either $f(t) = a + t^m$ and $\tilde{f}(t) = a + t^n$ or $f(t) = t^{-m}$ and $\tilde{f}(t) = t^{-n}$. Moreover

$$(s_k)^m = (t_k)^n.$$

Note also that if $s_k = t_k = 0$ for any k then $< f, g > = < \tilde{f}, \tilde{g} >$ which is what we want to prove. So we can assume that $s_k \neq 0$ and $t_k \neq 0$ for all k.

Let σ_k be any complex nth root of s_k. Then

$$\left(\frac{(\sigma_k)^m}{t_k}\right)^n = \frac{(s_k)^m}{(t_k)^n} = 1$$

so $\frac{(\sigma_k)^m}{t_k}$ is a complex nth root of 1 for each k. Since there are only n complex nth roots of 1 there must be at least one, ω say, such that

$$\frac{(\sigma_k)^m}{t_k} = \omega$$

for infinitely many values of k. For these values of k the equation

$$g(t^n) = \tilde{g}(\omega^{-1}t^m)$$

holds when $t = \sigma_k$. Since $\sigma_k \to 0$ as $k \to \infty$ and $\sigma_k \neq 0$ for all k it follows from the uniqueness theorem of complex analysis ([Priestley 85] 5.16) that

$$g(t^n) = \tilde{g}(\omega^{-1}t^m)$$

for all sufficiently small $t \in \mathbf{C}$. But also

$$f(t^n) = \tilde{f}(\omega^{-1}t^m)$$

since both sides are equal to $a + t^{mn}$ or t^{-mn}. Since the mappings

$$t \mapsto (f(t), g(t))$$

and

$$t \mapsto (\tilde{f}(t), \tilde{g}(t))$$

are one-to-one near 0, the mappings

$$t \mapsto (f(t^n), g(t^n))$$

and

$$t \mapsto (\tilde{f}(\omega^{-1}t^m), \tilde{g}(\omega^{-1}t^m))$$

are exactly n-to-one and m-to-one near 0. But we have just shown that these mappings are the same, so $m = n$. Thus

$$g(t^n) = \tilde{g}(\omega^{-1}t^n)$$

and hence

$$g(t) = \tilde{g}(\omega^{-1}t)$$

for all sufficiently small t. Also

$$f(t) = \tilde{f}(\omega^{-1}t)$$

so

$$(f(t), g(t)) = (\tilde{f}(\rho(t)), \tilde{g}(\rho(t)))$$

where $\rho(t) = \omega^{-1}t$ defines a parameter change. Thus

$$< f, g >=< \tilde{f}, \tilde{g} >$$

as required. $\square$

7.2 Newton polygons and Puiseux expansions

In this section we shall investigate what a projective curve C of degree d in $\mathbf{P}_2$ looks like near a singular point. To simplify notation we shall choose coordinates so that the point in question is

$$[0,0,1].$$

Then the curve is

$$C = \{[x,y,z] \in \mathbf{P}_2 \ : \ P(x,y,z) = 0\}$$

where $P(0,0,1) = 0$. We shall find (see theorem 7.28 and remark 7.29 below) that there are positive integers $m_1, \ldots, m_k$ and power series in x^{1/m_j} (called Puiseux expansions)

$$\sum_{r \geq 1} a_r^{(j)} x^{r/m_j}$$

for $1 \leq j \leq k$ such that if x and y are near 0 then the Puiseux expansions converge and

$$P(x,y,1) = 0$$

if and only if

$$y = \sum_{r \geq 1} a_r^{(j)} (x^{1/m_j})^r$$

for some $j \in \{1, \ldots, k\}$ and some choice of the m_jth root x^{1/m_j} of x. We shall also describe a method of calculating Puiseux expansions via the polygons used by Newton.

Newton's idea was to think of the equation

$$P(x,y,1) = 0$$

as an implicit equation for y as a function of x near 0, and then to expand y as a series of fractional powers of x. When

$$\frac{\partial P}{\partial y}(0,0,1) \neq 0$$

the implicit function theorem tells us that locally y is a holomorphic single-valued function of x and hence can be expanded as a power series in x. However the example of the cubic curve defined by

$$y^2 z = x^3$$

shows that in general we must allow fractional powers of x.

Let us suppose first that $P(x, y, 1)$ is a *quasi-homogeneous* polynomial, in the sense that there are positive rational numbers μ, ν such that

$$P(x, y, 1) = \sum_{\alpha + \mu\beta = \nu} c_{\alpha\beta} x^\alpha y^\beta$$

for suitable coefficients $c_{\alpha\beta}$. (It is easy to see that the polynomial $P(x, y, 1)$ is quasi-homogeneous if and only if there is a line in $\mathbf{R}^2$ containing all the points $(\alpha, \beta) \in \mathbf{Z}^2$ such that the coefficient of the monomial $x^\alpha y^\beta$ in $P(x, y, 1)$ is nonzero). Then substituting

$$y = tx^\mu$$

in $P(x, y, 1)$ we find

$$P(x, tx^\mu, 1) = \sum_{\alpha + \mu\beta = \nu} c_{\alpha\beta} x^\alpha t^\beta x^{\mu\beta} = x^\mu f(t)$$

where

$$f(t) = \sum_{\alpha + \mu\beta = \nu} c_{\alpha\beta} t^\beta$$

is a polynomial in t. We can always find $t \in \mathbf{C}$ such that $f(t) = 0$ and then $y = tx^\mu$ is a solution to the equation

$$P(x, y, 1) = 0.$$

In general let

$$P(x, y, 1) = \sum_{\alpha, \beta} c_{\alpha\beta} x^\alpha y^\beta$$

and define the *carrier* $\Delta(P)$ of $P(x, y, 1)$ to be

$$\Delta(P) = \{(\alpha, \beta) \in \mathbf{Z}^2 : c_{\alpha\beta} \neq 0\}.$$

Suppose that μ and ν are positive rational numbers such that

$$P(x, y, 1) = \sum_{\alpha + \mu\beta \geq \nu} c_{\alpha\beta} x^\alpha y^\beta;$$

i.e. $\Delta(P)$ lies in the half plane above and to the right of the line in $\mathbf{R}^2$ defined by

$$x + \mu y = \nu.$$

Suppose also that there are at least two points $(\alpha, \beta) \in \Delta(P)$ with $\alpha + \mu\beta = \nu$. If we substitute

$$y = tx^\mu$$

into $P(x, y, 1)$ we get

$$P(x, tx^\mu, 1) = x^\nu f(t) + \sum_{\alpha + \mu\beta > \nu} c_{\alpha\beta} t^\beta x^{\alpha + \mu\beta}$$

where

$$f(t) = \sum_{\alpha + \mu\beta = \nu} c_{\alpha\beta} t^\beta.$$

This time by choosing t to be a nonzero root of $f(t)$ we cannot necessarily find a solution to

$$P(x, y, 1) = 0$$

but we can at least make the lowest possible μ-order terms vanish, where the μ-order of a monomial $x^\alpha y^\beta$ is $\alpha + \mu\beta$. This means that we can regard

$$y = tx^\mu$$

as in some sense an approximate solution to $P(x, y, 1)$. In fact if we choose μ as small as possible then tx^μ will be the first term in our expansion for y as a series of fractional powers of x.

To show that this process can be made to work we first construct the *Newton polygon* of P.

Definition 7.25 *If $p, q \in \mathbf{R}^2$ let*

$$[p, q] = \{tp + (1 - t)q : 0 \le t \le 1\}$$

be the straight line segment from p to q. Consider the convex subset of $\mathbf{R}^2$ consisting of those $(x, y) \in \mathbf{R}^2$ such that

$$x \ge a \quad and \quad y \ge b$$

for some $(a, b) \in [\delta_1, \delta_2]$ where δ_1 and δ_2 belong to the carrier $\Delta(P)$. Its boundary consists of a vertical half-line and a horizontal half-line joined by a union of finitely many straight line segments. This union is called the Newton polygon *of P (see figure 7.1).*

Note that $(0, 0) \notin \Delta(P)$ since $P(0, 0, 1) = 0$. Also we can always choose coordinates so that $P(x, y, z)$ is not divisible by x; let us assume that this is done so there is some β such that

$$(0, \beta) \in \Delta(P).$$

Then if the Newton polygon is a single point it must be $(0, \beta_0)$ for some $\beta_0 > 0$ where

$$P(x, y, 1) = y^{\beta_0} Q(x, y)$$

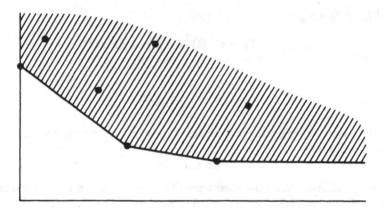

Figure 7.1: A Newton polygon

with $Q(0,0) \neq 0$. In this case the only solutions to

$$P(x,y,1) = 0$$

near $(0,0)$ are given by $y = 0$. Otherwise the steepest segment of the Newton polygon is the starting line for the procedure described above. Let $(0, \beta_0)$ be the upper endpoint of this segment and let $-\frac{1}{\mu_0}$ be its slope. Then μ_0 is a positive rational, say

$$\mu_0 = \frac{p_0}{q_0}$$

where p_0, q_0 are coprime positive integers. Moreover we can write

$$P(x,y,1) = \sum_{\alpha + \mu_0 \beta \geq \nu_0} c_{\alpha\beta} x^\alpha y^\beta$$

where

$$\nu_0 = \mu_0 \beta_0$$

and there is at least one point $(\alpha, \beta) \in \Delta(P)$ other than $(0, \beta_0)$ such that $\alpha + \mu_0 \beta = \nu_0$. It follows that the polynomial

$$f_0(t) = \sum_{\alpha + \mu_0 \beta = \nu_0} c_{\alpha\beta} t^\beta$$

has a nonzero root, t_0 say. Then

$$y_0 = t_0 x^{\mu_0}$$

gives us a first approximate solution to the equation

$$P(x,y,1) = 0.$$

Next we make the substitution

$$x = (x_1)^{q_0}$$

and

$$y = x^{\mu}(t_0 + y_1) = x_1^{p_0}(t_0 + y_1)$$

to get

$$
\begin{aligned}
P(x_1^{q_0}, x_1^{p_0}(t_0 + y_1), 1) &= \sum_{\alpha + \mu_0 \beta \geq \nu_0} c_{\alpha\beta} x_1^{q_0(\alpha + \mu_0\beta)}(t_0 + y_1)^{\beta} \\
&= x_1^{q_0\nu_0} P_1(x_1, y_1)
\end{aligned}
$$

where

$$P_1(x_1, y_1) = \sum_{q_0\alpha + p_0\beta \geq q_0\nu_0} c_{\alpha\beta} x_1^{q_0\alpha + p_0\beta - q_0\nu_0}(t_0 + y_1)^{\beta}$$

is a polynomial in x_1 and y_1, not divisible by x_1.

Remark 7.26 Recall that y^{β_0} is the smallest power of y occuring in $P(x, y, 1)$; let $y_1^{\beta_1}$ be the smallest power of y_1 occurring in $P(x_1, y_1)$. We can note at this point that

$$P(0, y_1) = \sum_{\alpha + \mu_0\beta = \nu_0} c_{\alpha\beta}(t_0 + y_1)^{\beta}$$

so β_1 is the smallest positive integer β satisfying $c_{\alpha\beta} \neq 0$ and $\alpha + \mu_0\beta = \nu_0$ for some $\alpha \geq 0$. Since $c_{0\beta_0} \neq 0$ and $\mu_0\beta_0 = \nu_0$ and $P_1(0,0) = 0$ we must have

$$\beta_1 < \beta_0$$

unless $P_1(0, y_1)$ is a constant multiple of $y_1^{\beta_0}$. If this is the case the term

$$c_{0\beta_0}\beta_0 t_0 y_1^{\beta_0 - 1}$$

in

$$c_{0\beta_0}(t_0 + y_1)^{\beta_0}$$

must cancel with

$$c_{\alpha,\beta_0 - 1} y_1^{\beta_0 - 1}$$

where

$$\alpha + \mu_0(\beta_0 - 1) = \nu_0 = \mu_0\beta_0$$

and hence

$$\mu_0 = \alpha$$

is an integer, or equivalently $q_0 = 1$. This will be important later.

We now repeat the whole process, replacing $P(x,y,1)$ by $P(x_1,y_1)$, and continue indefinitely. We obtain a sequence of positive rationals

$$\mu_0 = \frac{p_0}{q_0}, \; \mu_1 = \frac{p_1}{q_1}, \; \mu_2 = \frac{p_2}{q_2}, \ldots$$

and complex numbers

$$t_0, \; t_1, \; t_2, \ldots$$

and successive "approximate solutions"

$$(x,y), \; (x_1,y_1), \; (x_2,y_2), \ldots$$

to the equation

$$P(x,y,1) = 0$$

related by

$$x = x_1^{1/q_0}, \; x_1 = x_2^{1/q_1}, \; x_2 = x_3^{1/q_2}, \ldots$$

and

$$y = x^{\mu_0}(t_0 + y_1), \; y_1 = x_1^{\mu_1}(t_1 + y_2), \; y_2 = x_2^{\mu_2}(t_2 + y_3), \ldots.$$

We would like to conclude that

$$\begin{aligned} y &= t_0 x^{\mu_0} + t_1 x_1^{\mu_1} x^{\mu_0} + t_2 x_2^{\mu_2} x_1^{\mu_1} x^{\mu_0} + \cdots \\ &= t_0 x^{\mu_0} + t_1 x^{\mu_0 + \mu_1/q_0} + t_2 x^{\mu_0 + \mu_1/q_0 + \mu_2/q_0 q_1} + \cdots. \end{aligned}$$

is a genuine solution near $(0,0)$. This series is called a Puiseux expansion for the curve

$$C = \{[x,y,z] \in \mathbf{P}_2 \; : \; P(x,y,z) = 0\}$$

near $[0,0,1]$. Note that by remark 7.26 we have

$$q_i = 1$$

unless

$$\beta_{i-1} > \beta_i$$

where

$$\beta_0 \geq \beta_1 \geq \beta_2 \geq \cdots$$

is a decreasing sequence of positive integers. Hence there are at most finitely many i such that $q_i > 1$. Let n be the product of the q_i for these values of i; then the Puiseux expansion may be expressed as a formal power series

$$y = \sum_{r \geq 1} a_r x^{r/n}$$

in $x^{1/n}$. Before showing that this series does converge for small x and defines a solution to the equation

$$P(x,y,1) = 0$$

let us illustrate the process with an example.

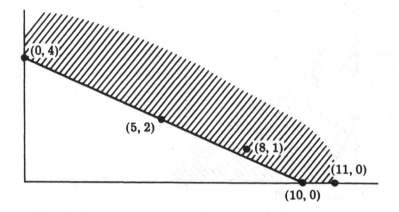

Figure 7.2: The Newton polygon of $y^4z^7 - 2x^5y^2z^4 - 4x^8yz^2 + x^{10}z - x^{11}$

Example 7.27 Let

$$P(x, y, z) = y^4z^7 - 2x^5y^2z^4 - 4x^8yz^2 + x^{10}z - x^{11}$$

so that $P(x, y, z)$ is homogeneous of degree 11 and

$$P(x, y, 1) = y^4 - 2x^5y^2 - 4x^8y + x^{10} - x^{11}.$$

In this case the Newton polygon consists of a single line segment (see figure 7.2). We take $\mu_0 = 5/2$ and $\nu_0 = 10$. Then the polynomial

$$f_0(t) = \sum_{\alpha + \mu_0\beta = \nu_0} c_{\alpha\beta}t^\beta = 1 - 2t^2 + t^4 = (1 - t^2)^2$$

has roots $t_0 = \pm 1$. Thus our first approximate solutions are $y = \pm x^{5/2}$. Substituting

$$x = x_1^2$$

and

$$y = x_1^5(\pm 1 + y_1)$$

in $P(x, y, 1)$ and taking out the factor x_1^{20} gives

$$
\begin{aligned}
P_1(x_1, y_1) &= (\pm 1 + y_1)^4 - 2(\pm 1 + y_1)^2 - 4x_1(\pm 1 + y_1) + 1 - x_1^2 \\
&= y_1^4 \pm 4y_1^3 + 4y_1^2 - 4x_1y_1 - x_1^2 \mp 4x_1.
\end{aligned}
$$

Again the Newton polygon is a single line segment (figure 7.3). We take

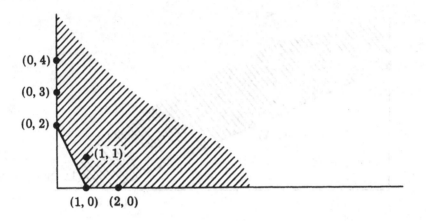

Figure 7.3: The Newton polygon of $y^4 \pm 4y^3z + 4y^2z^2 - 4xyz^2 - x^2z^2 \mp 4xz^3$

$\mu_1 = 1/2$ and $\nu_1 = 1$, and our next approximate solution is $y_1 = t_1 x_1^{1/2}$ where t_1 is a root of
$$4t^2 \mp 4 = 0,$$
i.e. $t_1 = \pm 1$ when $t_0 = 1$ and $t_1 = \pm i$ when $t_0 = -1$.

In each case it is easy to check that the approximate solution $y_1 = t_1 x_1^{1/2}$ is in fact a genuine solution to the equation

$$P_1(x_1, y_1) = 0$$

so the procedure stops at this point and the Puiseux expansions

$$y = x^{5/2} \pm x^{11/4}$$

and

$$y = -x^{5/2} \pm i x^{11/4}$$

are solutions to the equation

$$P(x, y, 1) = 0.$$

It is not in general the case that the Puiseux expansion terminates after finitely many terms. Nevertheless we have the following result.

Theorem 7.28 *Any Puiseux expansion*

$$y = \sum_{r \geq 1} a_r x^{r/n}$$

for the curve C near the point $[0,0,1]$ is a power series in $x^{1/n}$ which converges for x sufficiently close to 0 and satisfies

$$P(x, \sum_{r \geq 1} a_r x^{r/n}, 1) = 0.$$

Proof. By remark 7.19 there are holomorphic functions $g_1, \ldots, g_l$ defined near 0 and positive integers $m_1, \ldots m_l$ such that $m_1 + \cdots + m_l = d$ and a nonzero constant K such that

$$P(x, y, 1) = K \prod_{1 \leq j \leq l} \prod_{1 \leq s \leq m_j} (y - g_j(e^{2\pi i s/m_j} x^{1/m_j}))$$

for all y and all sufficiently small x. We can expand each $g_j(t)$ as a convergent power series

$$g_j(t) = \sum_{r \geq 0} a_r^{(j)} t^r$$

near 0.

Let N be the lowest common multiple of $m_1, \ldots, m_l$ and n. Then the series

$$g_j(e^{2\pi i s/m_j} x^{1/m_j}) = \sum_{r \geq 0} a_r^{(j)} e^{2\pi i r s/m_j} x^{r/m_j}$$

and the Puiseux expansion

$$\sum_{r \geq 1} a_r x^{r/n}$$

can all be regarded as elements of the ring $\mathbf{C}\{x^{1/N}\}$ of formal power series[1] in $x^{1/N}$. This ring is an integral domain, so if $Q(y)$ is a polynomial in y with coefficients in $\mathbf{C}\{x^{1/N}\}$ which satisfies $Q(c) = 0$ for some $c \in \mathbf{C}\{x^{1/N}\}$ and can be expressed in the form

$$Q(y) = K(y - c_1) \ldots (y - c_d)$$

for some $K \in \mathbf{C} - \{0\}$ and $c_1, \ldots, c_d \in \mathbf{C}\{x^{1/N}\}$, then

$$c = c_j$$

for some $j \in \{1, \ldots, d\}$. Therefore it suffices to show that, as a formal power series in $x^{1/N}$, the Puiseux expansion satisfies

$$P(x, \sum_{r \geq 1} a_r x^{r/n}, 1) = 0.$$

For then the Puiseux expansion must coincide with one of the series

$$\sum_{r \geq 0} a_r^{(j)} e^{2\pi i r s/m_j} x^{r/m_j}$$

---

[1] We add and multiply formal power series in the obvious way.

and hence must converge for sufficiently small x.

The construction of the Puiseux expansion shows that the exponent of the smallest power of $x^{1/N}$ occurring in the polynomial

$$P(x, \sum_{r=1}^{M} a_r x^{r/n}, 1)$$

is at least $p_0\beta_0 + p_1\beta_1 + \cdots + p_M\beta_M$, which tends to infinity as M tends to infinity since each p_j and β_j is a positive integer. This tells us that every coefficient in the formal power series

$$P(x, \sum_{r\geq 1} a_r x^{r/n}, 1)$$

is zero, i.e. that

$$P(x, \sum_{r\geq 1} a_r x^{r/n}, 1)$$

is zero in $\mathbf{C}\{x^{1/N}\}$ as required. $\square$

Remark 7.29 The proof of theorem 7.28 shows that there are holomorphic functions

$$g_j(t) = \sum_{r\geq 0} a_r^{(j)} t^r$$

for $1 \leq j \leq l$ defined near 0 and positive integers $m_1, \ldots, m_l$ such that $m_1 + \cdots + m_l = d$ so that the curve

$$C = \{[x, y, z] \in \mathbf{P}_2 \ : \ P(x, y, z) = 0\}$$

is given in a neighbourhood of the point $[0, 0, 1]$ by

$$\bigcup_{1\leq j\leq l, a_0^{(j)}=0} \ \bigcup_{1\leq s\leq m_j} \{[x, y, 1] \in \mathbf{P}_2 \ : \ y = g_j(e^{2\pi i s/m_j} x^{1/m_j})\},$$

and moreover any Puiseux expansion for C near $[0, 0, 1]$ is of the form

$$y = g_j(x^{1/m_j})$$

for some j such that $a_0^{(j)} = 0$. Conversely if $a_0^{(j)} = 0$ then by making appropriate choices in the construction of a Puiseux expansion we obtain the expansion

$$y = g_j(x^{1/m_j}).$$

It follows that calculating the Puiseux expansions of C near $[0, 0, 1]$ gives us a good picture of what C looks like near $[0, 0, 1]$.

Remark 7.30 Recall from theorem 7.12 that there is a compact connected Riemann surface $\tilde{C}$ and a continuous surjection $\pi : \tilde{C} \to C$ which restricts to a holomorphic bijection

$$\pi : \tilde{C} - \pi^{-1}(Sing(C)) \to C - Sing(C).$$

If we continue to use the notation of the last remark then it follows from remark 7.18 that

$$\pi^{-1}\{[0,0,1]\} = \{<t^{m_j}, g_j(t)> \,:\, 1 \le j \le l,\ g_j(0) = 0\}$$

where the Puiseux expansions of C near $[0,0,1]$ are given by

$$y = g_j(e^{2\pi i s/m_j} x^{1/m_j})$$

for $1 \le j \le l, 1 \le s \le m_j$ and $g_j(0) = 0$. We regard two such Puiseux expansions indexed by (j,s) and $(\tilde{j}, \tilde{s})$ as essentially different if $j \ne \tilde{j}$. Then the number of points in the inverse image of $[0,0,1]$ under $\pi : \tilde{C} \to C$ is the number of essentially different Puiseux expansions of C near $[0,0,1]$. Of course the same is true for any other point of C; we simply chose coordinates so that the point under consideration was $[0,0,1]$ for convenience.

7.3 The topology of singular curves

In Chapter 4 we gave a simple description of the topology of a *nonsingular* projective curve of degree d in $\mathbf{P}_2$: the curve is homeomorphic to a sphere with

$$g = \frac{1}{2}(d-1)(d-2)$$

handles. In this section we shall discuss the topology of singular projective curves in $\mathbf{P}_2$. We shall consider only irreducible curves; recall however that a reducible projective curve in $\mathbf{P}_2$ is the union of finitely many irreducible curves meeting at finitely many points.

Let

$$C = \{[x,y,z] \in \mathbf{P}_2 \,:\, P(x,y,z) = 0\}$$

be an irreducible curve of degree d in $\mathbf{P}_2$. We have seen (theorem 7.12) that there is a compact connected Riemann surface $\tilde{C}$ and a continuous surjection $\pi : \tilde{C} \to C$ which restricts to a homeomorphism

$$\pi : \tilde{C} - \pi^{-1}(Sing(C)) \to C - Sing(C).$$

Moreover if p is any singular point of C then the number of points in $\pi^{-1}\{p\}$ is determined by the Puiseux expansions of C near p (see remark 7.30). Thus topologically C is the result of identifying together certain finite sets of points in $\tilde{C}$ (see figure 4.2), and the way these identifications are made is determined by the Puiseux expansions of C near its singular points.

Topologically the compact connected Riemann surface $\tilde{C}$ is a sphere with g handles where g is the genus of $\tilde{C}$. We also call g the genus of the curve C. The aim of this section is to generalise the degree-genus formula for nonsingular curves to relate the degree and genus of our arbitrary irreducible projective curve C in $\mathbf{P}_2$. However the generalised formula also involves data concerning the singular points of C. We shall assign to each singular point p a positive integer $\delta(p)$ such that the formula

$$g = \frac{1}{2}(d-1)(d-2) - \sum_{p \in Sing(C)} \delta(p),$$

called *Noether's formula*, is true.

In order to prove Noether's formula we shall modify the proof given in §4.3 of the degree-genus formula for nonsingular curves. We assume that coordinates have been chosen so that

$$[0,1,0] \notin C$$

and we define $\phi : C \to \mathbf{P}_1$ by

$$\phi([x,y,z]) = [x,z].$$

We regard the composition

$$\psi = \phi \circ \pi : \tilde{C} \to \mathbf{P}_1$$

as a branched cover of $\mathbf{P}_1$. We define

$$R = \pi^{-1}\{[a,b,c] \in C \ : \ \frac{\partial P}{\partial y}(a,b,c) = 0\}$$

to be the set of *ramification points* of ψ, and call its image $\psi(R)$ the *branch locus* of ψ.

Proposition 7.31 *Given any triangulation* (V, E, F) *of* $\mathbf{P}_1$ *such that the branch locus* $\psi(R)$ *of* ψ *is contained in the set of vertices* V, *there is a triangulation* $(\tilde{V}, \tilde{E}, \tilde{F})$ *of* $\tilde{C}$ *such that*

$$\tilde{V} = \psi^{-1}(V)$$
$$\#\tilde{E} = d\#E$$

and

$$\#\tilde{F} = d\#F.$$

Proof. Since

$$R \supseteq \pi^{-1}(Sing(C))$$

and

$$\pi : \tilde{C} - \pi^{-1}(Sing(C)) \to C - Sing(C)$$

is a homeomorphism, this is a straightforward modification of the proof of proposition 4.22. □

As in §4.2 if $p = [a, b, c] \in C$ we define $\nu_\phi(p)$ to be the multiplicity of the zero of the polynomial $P(a, y, c)$ at $y = b$.

Lemma 7.32 *In the situation of proposition 7.31*

$$\#\tilde{V} = d\#V - \sum_{p \in \pi(R)} (\nu_\phi(p) - 1) + \sum_{p \in Sing(C)} (\#\pi^{-1}\{p\} - 1).$$

Proof. By lemma 4.5, which does not require C to be nonsingular, the inverse image under $\phi : C \to \mathbf{P}_1$ of any $q \in \mathbf{P}_1$ contains exactly

$$d - \sum_{p \in \phi^{-1}\{q\}} (\nu_\phi(p) - 1)$$

points. Moreover $\nu_\phi(p) = 1$ if $p \notin \pi(R)$ and $\phi^{-1}(V) \supseteq \pi(R)$. Thus

$$\#\phi^{-1}(V) = d\#V - \sum_{p \in \pi(R)} (\nu_\phi(p) - 1).$$

Since

$$\pi : \tilde{C} - \pi^{-1}(Sing(C)) \to C - Sing(C)$$

is a bijection and $\phi^{-1}(V)$ contains $Sing(C)$ it follows that

$$\begin{aligned} \#\psi^{-1}(V) &= \#\pi^{-1}\phi^{-1}(V) \\ &= d\#V - \sum_{p \in \pi(R)} (\nu_\phi(p) - 1) + \sum_{p \in Sing(C)} (\#\pi^{-1}\{p\} - 1) \end{aligned}$$

as required. □

Let us write $I_p(P, \frac{\partial P}{\partial y})$ for the intersection multiplicity at p of the curve C defined by $P(x, y, z)$ and the curve in $\mathbf{P}_2$ defined by the homogeneous polynomial $\frac{\partial P}{\partial y}(x, y, z)$ of degree $d - 1$.

Lemma 7.33 *Suppose that coordinates are chosen on $\mathbf{P}_2$ so that $[0, 1, 0]$ does not lie on C or the tangent line to C at any of the finitely many points $p \in C - Sing(C)$ which are inflection points on C. Then if $p \in \pi(R)$ and $p \notin Sing(C)$ we have*

$$\nu_\phi(p) = 2$$

and

$$I_p(P, \frac{\partial P}{\partial y}) = 1.$$

Proof. This follows from the proofs of lemmas 4.8 and 4.7 and proposition 3.22. □

Corollary 7.34 *If coordinates are chosen as in lemma 7.33 then the Euler number $\chi(\tilde{C})$ of $\tilde{C}$ is given by*

$$\chi(\tilde{C}) = d(3-d) + \sum_{p \in Sing(C)} (I_p(P, \frac{\partial P}{\partial y}) - \nu_\phi(p) + \#\pi^{-1}\{p\}).$$

Proof. By definition

$$\chi(\tilde{C}) = \#\tilde{V} - \#\tilde{E} + \#\tilde{F}$$

where $(\tilde{V}, \tilde{E}, \tilde{F})$ is any triangulation of $\tilde{C}$. Therefore by proposition 7.31 and lemma 7.32

$$\chi(\tilde{C}) = d(\#V - \#E + \#F) - \sum_{p \in \pi(R)} (\nu_\phi(p) - 1) + \sum_{p \in Sing(C)} (\#\pi^{-1}\{p\} - 1)$$

where (V, E, F) is a triangulation of $\mathbf{P}_1$. Since

$$\chi(\mathbf{P}_1) = 2$$

by 4.14(i) we have

$$\#V - \#E + \#F = 2.$$

Moreover by lemma 7.33

$$\sum_{p \in \pi(R) - Sing(C)} (\nu_\phi(p) - 1) \doteq \sum_{p \in \pi(R) - Sing(C)} I_p(P, \frac{\partial P}{\partial y}).$$

Since $\pi(R)$ is the intersection of the curve C in $\mathbf{P}_2$ defined by $P(x, y, z)$ and the curve in $\mathbf{P}_2$ defined by $\frac{\partial P}{\partial y}(x, y, z)$ and

$$Sing(C) \subseteq \pi(R)$$

it follows from Bézout's theorem (theorem 3.1) that

$$\sum_{p \in \pi(R) - Sing(C)} I_p(P, \frac{\partial P}{\partial y}) = d(d-1) - \sum_{p \in Sing(C)} I_p(P, \frac{\partial P}{\partial y}).$$

Combining these equalities gives the required formula for $\chi(\tilde{C})$. □

Definition 7.35 *Let p be a singular point of the irreducible curve*

$$C = \{[x, y, z] \in \mathbf{P}_2 \ : \ P(x, y, z) = 0\}$$

and suppose that coordinates have been chosen so that $[0, 1, 0]$ does not lie on C or on the tangent line to C at any of the finitely many points $p \in C - Sing(C)$ which are inflection points on C. We define

$$\delta(p) = \frac{1}{2}(I_p(P, \frac{\partial P}{\partial y}) - \nu_\phi(p) + \#\pi^{-1}\{p\}).$$

Remark 7.36 Although we shall not prove this fact here, it can be shown that $\delta(p)$ is always a positive integer (cf. [Brieskorn & Knörrer 86] §9.2).

We shall show that the definition of $\delta(p)$ is independent of the choice of coordinates. Then we will have Noether's formula.

Theorem 7.37 (Noether's formula.) *The genus g of an irreducible projective curve C of degree d in $\mathbf{P_2}$ is*

$$g = \frac{1}{2}(d-1)(d-2) - \sum_{p \in Sing(C)} \delta(p).$$

Proof. This follows immediately from corollary 7.34 because

$$\chi(\tilde{C}) = 2 - 2g.$$

Examples 7.38
(i) Consider the cuspidal cubic curve C defined by the polynomial

$$P(x, y, z) = y^2 z - x^3.$$

C has one singular point $p = [0, 0, 1]$ and its Puiseux expansions near $[0, 0, 1]$ are given by

$$y = \pm x^{3/2}.$$

These are not essentially different, so

$$\#\pi^{-1}\{p\} = 1$$

(see remark 7.30). Moreover

$$P(0, y, 1) = y^2$$

has a zero of multiplicity two at $y = 0$ so $\nu_\phi(p) = 2$. Finally

$$\frac{\partial P}{\partial y}(x, y, z) = 2yz$$

so by theorem 3.18(i),(iii),(iv) and (v)

$$
\begin{aligned}
I_p(P, \frac{\partial P}{\partial y}) &= I_p(P, y) + I_p(P, z) \\
&= I_p(x^3, y) + 0 \\
&= 3I_p(x, y) \\
&= 3.
\end{aligned}
$$

Thus

$$\delta(p) = \frac{1}{2}(3 - 2 + 1) = 1$$

so the genus of C is

$$g = \frac{1}{2}(3 - 1)(3 - 2) - 1 = 0.$$

Thus the resolution of singularities $\tilde{C}$ of C is topologically a sphere, and so is C since $\pi : \tilde{C} \to C$ is a homeomorphism.

(ii) Now consider the nodal cubic curve C defined by the polynomial

$$P(x, y, z) = y^2 z - x^3 - x^2 z.$$

Again C has one singular point $p = [0, 0, 1]$. Its Puiseux expansions near $[0, 0, 1]$ are

$$y = \pm x(1 + \frac{1}{2}x - \frac{1}{8}x^2 + \cdots).$$

These are essentially different in the sense of remark 7.30 so

$$\#\pi^{-1}\{p\} = 2.$$

Also

$$P(0, y, 1) = y^2$$

has a zero of multiplicity two at $y = 0$ so $\nu_\phi(p) = 2$. Finally by theorem 3.18(i),(iii),(iv) and (v)

$$
\begin{aligned}
I_p(P, \frac{\partial P}{\partial y}) &= I_p(P, y) + I_p(P, z) \\
&= I_p(x^2(x + z), y) + 0 \\
&= 2I_p(x, y) + I_p(x + z, y) \\
&= 2.
\end{aligned}
$$

Thus

$$\delta(p) = \frac{1}{2}(2 - 2 + 2) = 1$$

so the genus of C is 0 as before. Hence $\tilde{C}$ is a sphere but this time $\pi : \tilde{C} \to C$ is not a homeomorphism; topologically C is a sphere with two points identified (cf. figure 1.1).

(iii) The singular point of a nodal cubic curve is an ordinary double point. In fact if p is any ordinary singular point of multiplicity m on a curve C then

$$\delta(p) = \frac{1}{2}m(m - 1).$$

This is because

$$\#\pi^{-1}\{p\} = m$$

and if we choose coordinates appropriately then $\nu_\phi(p) = m$ and

$$I_p(P, \frac{\partial P}{\partial y}) = m(m-1)$$

by the remark following proposition 3.22.

In particular it follows that if C is an irreducible curve of degree d in $\mathbf{P}_2$ with r ordinary double points and no other singular points, then topologically C is a sphere with

$$\frac{1}{2}(d-1)(d-2) - r$$

handles with $2r$ points identified together in pairs.

We now just have to show that the definition of $\delta(p)$ is independent of the choice of coordinates.

It is clear that $\delta(p)$ is unaffected by changes of coordinates that respect the operation $\partial/\partial y$ (up to multiplication by a nonzero scalar) since intersection multiplicities are preserved by projective transformations. This means that it is enough to prove the following lemma.

Lemma 7.39 *Suppose that neither* $[0,1,0]$ *nor* $[\alpha,\beta,\gamma]$ *lie on* C *or the tangent line to* C *at any inflection point on* $C - Sing(C)$. *Suppose that* $p \in Sing(C)$ *and let* $\nu_\phi^{[\alpha,\beta,\gamma]}(p)$ *be the smallest positive integer such that*

$$(\alpha\frac{\partial}{\partial x} + \beta\frac{\partial}{\partial y} + \gamma\frac{\partial}{\partial z})^m P$$

does not vanish at p. *Then*

$$I_p(P, \alpha\frac{\partial P}{\partial x} + \beta\frac{\partial P}{\partial y} + \gamma\frac{\partial P}{\partial z}) - \nu_\phi^{[\alpha,\beta,\gamma]}(p) = I_p(P, \frac{\partial P}{\partial y}) - \nu_\phi(p).$$

Proof. If $[\alpha,\beta,\gamma] = [0,1,0]$ we have nothing to prove. Otherwise we can find a projective transformation which preserves $[0,1,0]$ and respects the operation $\partial/\partial y$ and takes $[\alpha,\beta,\gamma]$ to $[1,0,0]$. Thus we can assume that $[\alpha,\beta,\gamma] = [1,0,0]$.

If $h(t)$ is a nonconstant meromorphic function defined on a connected open neighbourhood of 0 in $\mathbf{C}$ let $\mu(h)$ be the multiplicity of the zero or minus the multiplicity of the pole of h at 0.

Lemma 7.40 *(i) We have*

$$I_p(P, \frac{\partial P}{\partial y}) = \sum_{<f,g> \in \pi^{-1}\{p\}} \mu(\frac{\partial P}{\partial y}(f,g,1))$$

where

$$\mu(\frac{\partial P}{\partial y}(f,g,1))$$

is the multiplicity of the zero or minus the multiplicity of the pole of

$$\frac{\partial P}{\partial y}(f(t),g(t),1)$$

at $t = 0$.
(ii) We also have

$$\nu_\phi(p) = \sum_{<f,g> \in \pi^{-1}\{p\}} \mu(f - f(0))$$

where $f(0)$ *is omitted if it is equal to* ∞.

Given this lemma we can complete the proof of lemma 7.39. For if $< f,g > \in \tilde{C}$ then
$$P(f(t),g(t),1) = 0$$
for all t near 0, and differentiating this equation gives

$$f'(t)\frac{\partial P}{\partial x}(f(t),g(t),1) + g'(t)\frac{\partial P}{\partial y}(f(t),g(t),1) = 0$$

for all t near 0. Thus

$$\mu(f') + \mu(\frac{\partial P}{\partial x}(f,g,1)) = \mu(g') + \mu(\frac{\partial P}{\partial y}(f,g,1))$$

Since
$$\mu(f') = \mu(f - f(0)) - 1$$
and
$$\mu(g') = \mu(g - g(0)) - 1,$$

where $f(0)$ and $g(0)$ are omitted if equal to ∞, summing this equality over all $< f,g > \in \tilde{C}$ such that $\pi(< f,g >) = p$ and using (7.40) as it stands and with the roles of x and y reversed gives

$$I_p(P, \frac{\partial P}{\partial x}) + \nu_\phi(p) = I_p(P, \frac{\partial P}{\partial y}) + \nu_\phi^{[1,0,0]}(p).$$

The result follows. $\square$

Finally we prove lemma 7.40.

Proof of lemma 7.40. (i). By remark 7.19 if $p = [a, b, c]$ with $c \neq 0$ we can assume that $c = 1$ and write

$$P(x, y, z) = K \prod_{1 \leq j \leq l} \prod_{1 \leq s \leq m_j} (y - zg_j(e^{2\pi i s/m_j}(\frac{x}{z} - a)^{1/m_j}))$$

where K is a nonzero constant and $g_1, \ldots, g_l$ are holomorphic functions defined near $a \in \mathbb{C}$, and

$$\pi^{-1}\{p\} = \{< a + t^{m_j}, g_j(t) > :\ 1 \leq j \leq l, g_j(0) = b\}.$$

If $c = 0$ then we replace $a + t^{m_j}$ by t^{-m_j} and

$$zg_j(e^{2\pi i s/m_j}(\frac{x}{z} - a)^{1/m_j})$$

by

$$xg_j(e^{2\pi i s/m_j}(\frac{z}{x})^{1/m_j}).$$

Even though

$$y - zg_j(e^{2\pi i s/m_j}(\frac{x}{z} - a)^{1/m_j})$$

is not a polynomial, the proof of theorem 3.18 can be modified to show that the equations

$$
\begin{aligned}
I_p(P, \frac{\partial P}{\partial y}) &= \sum_{1 \leq j \leq l} \sum_{1 \leq s \leq m_j} I_p(y - zg_j(e^{2\pi i s/m_j}(\frac{x}{z} - a)^{1/m_j}), \frac{\partial P}{\partial y}) \\
&= \sum_{1 \leq j \leq l, g_j(0)=b} \sum_{1 \leq s \leq m_j} \mu(\frac{\partial P}{\partial y}(a + t, g_j(e^{2\pi i s/m_j} t^{1/m_j}), 1)) \\
&= \sum_{1 \leq j \leq l, g_j(0)=b} \mu(\frac{\partial P}{\partial y}(a + t^{m_j}, g_j(t), 1)) \\
&= \sum_{<f,g> \in \pi^{-1}\{p\}} \mu(\frac{\partial P}{\partial y}(f, g, 1))
\end{aligned}
$$

make sense and are valid (with appropriate modifications if $c = 0$).

(ii). Continuing the argument of (i) we find that the multiplicity $\nu_\phi(p)$ of the zero of $P(a, y, c)$ at $y = b$ is

$$\sum_{1 \leq j \leq l, g_j(0)=b} m_j = \sum_{<f,g> \in \pi^{-1}\{p\}} \mu(f - f(0))$$

as required (appropriate modifications having been made if $c = 0$). $\square$

7.4   Exercises

7.1. Let g be a meromorphic function defined near 0 in $\mathbf{C}$ with Laurent expansion

$$g(t) = \sum_{k=-\infty}^{\infty} a_k t^k.$$

Suppose that no positive integer except 1 is a common factor of all those $k \in \mathbf{Z}$ such that $a_k \neq 0$. Show that if m and n are coprime integers then

$$(t^m, g(t^n))$$

is a pair.
[Hint: compare the proof of 7.25.]

7.2. Let S be a connected Riemann surface and let f and g be meromorphic functions on S such that f is not constant. Show that there is a unique holomorphic map $h : S \to \mathcal{M}$ such that

$$f = \psi \circ h$$

and

$$g = \chi \circ h$$

where ψ and χ are defined as at remark 7.7. Show that if S is compact the image of h is a connected component of $\mathcal{M}$.

7.3. Let S be a compact connected Riemann surface. Assume that if p, q are distinct points of S then there is a meromorphic function f on S such that $f(p) \neq f(q)$. [This is in fact always true: see e.g. [Jones 71] Chapter VI.]
(i). Use exercise 5.17 to show that there is a nonconstant meromorphic function f on S taking each value in $\mathbf{P}_1$ exactly n times, counting multiplicities, for some $n \geq 1$.
(ii). Show that there is a meromorphic function g on S and some $a \in \mathbf{C}$ such that

$$\{g(p) \; : \; f(p) = a\}$$

has n distinct points, as follows. If $n = 1$ take $g = 0$. Otherwise show there is some $a \in \mathbf{C}$ such that

$$f^{-1}\{a\} = \{p_1, \ldots, p_n\}$$

where $p_1, \ldots, p_n$ are distinct. For each $j > 1$ choose a meromorphic function h_j on S such that

$$h_j(p_1) \neq h_j(p_j).$$

By replacing h_j by

$$p \mapsto \frac{\alpha h_j(p) + \beta}{\gamma h_j(p) + \delta}$$

for suitable $\alpha, \beta, \gamma, \delta$ show that we can assume that

$$h_j(p_1) = 1, \quad h_j(p_j) = 0$$

and

$$h_j(p_k) \neq \infty$$

if $1 \leq k \leq n$. Let

$$g_1 = \prod_{j=2}^{n} h_j$$

and show that $g_1(p_1) = 1$ and $g_1(p_j) = 0$ if $2 \leq j \leq n$. Choose $g_2, \ldots, g_n$ similarly and show that

$$g = \sum_{k=1}^{n} k g_k$$

has the required property.

(iii). Now show that the unique holomorphic map $h : S \to \mathcal{M}$ such that $f = \psi \circ h$ and $g = \chi \circ h$ given by exercise 7.2 is a holomorphic bijection onto a connected component of $\mathcal{M}$. [To show that h is injective it suffices by exercise 5.17 to show that there is some $q \in \mathcal{M}$ such that h takes the value q exactly once, counting multiplicities. Take $q = h(p_1)$ where $f(p_1) = a$ and a is chosen as in (ii).]

7.4. Find the first few terms of the Puiseux expansions about $[0, 0, 1]$ of the curves defined by
(i) $y^4 z^3 + 2x^3 y^2 z^2 + 4x^5 yz + x^6 z + x^7$;
(ii) $x^3 + y^3 + 3xyz$;
(iii) $2xy^5 + 5y^2 z^4 - 3x^2 z^4$.

7.5. Find the singular points, calculate the first few terms of the Puiseux expansions about the singular points and use Noether's formula to find the genus of the curves defined by
(i) $y^2 z = x^3$;
(ii) $y^2 z = x^2 (x + z)$;
(iii) $x^3 + y^3 + 3xyz$;
(iv) $2xy^5 + 5y^2 z^4 - 3x^2 z^4$.

7.6. Let C and D be projective curves in $\mathbf{P}_2$ defined by polynomials $P(x, y, z)$ and $Q(x, y, z)$. Suppose that C has Puiseux expansions $g_j(t^{1/k_j})$ about $[0, 0, 1]$ for $1 \leq j \leq l$. Show that $I_{[0,0,1]}(C, D)$ is the sum over $j \in \{1, \ldots, l\}$ of the multiplicities of the zeros of the functions

$$t \mapsto Q(t^{k_j}, g_j(t), 1)$$

at 0 (cf. the proof of 7.40(i)).

7.7. (i). Recall that the projective curve in $\mathbf{P}_2$ defined by

$$y^2 z^{k-2} = (x - \alpha_1 z) \ldots (x - \alpha_k z),$$

where $\alpha_1, \ldots, \alpha_k$ are distinct complex numbers, is called an *elliptic* curve if its degree k is 3 or 4, and a *hyperelliptic* curve if its degree k is at least 5. Show that if $k \le 3$ then the curve is nonsingular and that if $k > 3$ then it has a unique singular point $[0, 1, 0]$.

(ii). Let g be a positive integer and let $\alpha_1, \ldots, \alpha_{2g+1}$ be distinct complex numbers. Use the implicit function theorem of Appendix B to show that there are open neighbourhoods V and W of 0 in $\mathbf{C}$ and a holomorphic function $f : V \to W$ such that $f(0) = 0$ and if $x \in W$ and $y \in V$ then

$$x = y^2 (1 - \alpha_1 x) \ldots (1 - \alpha_{2g+1} x) \Leftrightarrow x = f(y).$$

(iii). Let C be the projective curve defined by

$$y^2 z^{2g-1} = (x - \alpha_1 z) \ldots (x - \alpha_{2g+1} z).$$

Show that if $x \ne 0 \ne y$ then

$$y^2 = (x - \alpha_1) \ldots (x - \alpha_{2g+1})$$

if and only if

$$\frac{1}{x} = \left(\frac{x^g}{y} \right)^2 \left(1 - \frac{\alpha_1}{x} \right) \ldots \left(1 - \frac{\alpha_{2g+1}}{x} \right).$$

Writing $w = \frac{x^g}{y}$ deduce that if $f : V \to W$ is as in (ii) the map from the open subset V of $\mathbf{C}$ to C defined by

$$w \mapsto [w(f(w))^{g-1}, 1, w(f(w))^g]$$

is a homeomorphism of V onto an open neighbourhood U of $[0, 1, 0]$ in C, with inverse $\phi : U \to V$ given by

$$\phi[x, y, z] = \begin{cases} x^g / y z^{g-1} & \text{if } [x, y, z] \ne [0, 1, 0], \\ 0 & \text{if } [x, y, z] = [0, 1, 0]. \end{cases}$$

(iv). Show that the restriction to $U - \{[0, 1, 0]\}$ of the standard holomorphic atlas Φ on $C - Sing(C)$ is compatible with the restriction of the holomorphic atlas $\Psi = \{\phi : U \to V\}$ on U. (It suffices to show that $\phi : U \to V$ is holomorphic with respect to Φ).

Deduce from exercise 5.8 that $\Phi \cup \Psi$ is a holomorphic atlas on C, and hence makes C into a Riemann surface.

(v). Show that the Riemann surface $\tilde{C}$ of the multivalued function

$$\sqrt{(x - \alpha_1)(x - \alpha_2)\ldots(x - \alpha_{2g+1})}$$

of x contains a unique pair mapping to the point $[0, 1, 0]$ of C, which is the pair

$$< \frac{1}{f(t)}, \frac{1}{tf(t)^g} > .$$

Deduce that the resolution of singularities $\pi : \tilde{C} \to C$ is a homeomorphism, which is holomorphic with respect to the atlas $\Phi \cup \Psi$ on C.

(vi). C is homeomorphic to a sphere with g handles (see §1.2.3). Deduce that there exist compact connected Riemann surfaces of every genus $g \geq 0$ and use the degree-genus formula to show that not every compact connected Riemann surface is biholomorphic to a nonsingular projective curve in $\mathbf{P}_2$. [However it is in fact the case that every compact connected Riemann surface is biholomorphic to the resolution of singularities $\tilde{C}$ of some (possibly singular) projective curve in $\mathbf{P}_2$; this is proved in, for example, [Jones 71].]

(vii). Use (vi) and Noether's formula or a direct calculation using exercise 7.6 to show that

$$\delta_{[0,1,0]} = 2g(g - 1).$$

7.8. (i). Let C be the hyperelliptic curve in $\mathbf{P}_2$ defined by

$$y^2 z^{2g-2} = (x - \alpha_1 z)\ldots(x - \alpha_{2g} z)$$

where $\alpha_1, \ldots, \alpha_{2g}$ are distinct complex numbers. Show that $y \to \infty$ as $[x, y, 1] \in C$ tends to the singular point $[0, 1, 0]$ and deduce from the equation of C that $x \to \infty$ also.

Show that if R is sufficiently large the holomorphic function

$$(1 - \frac{\alpha_1}{x})\ldots(1 - \frac{\alpha_{2g}}{x})$$

of $x \in \mathbf{C} - \{0\}$ has two holomorphic square-roots,

$$\pm\sqrt{(1 - \frac{\alpha_1}{x})\ldots(1 - \frac{\alpha_{2g}}{x})},$$

say, in $\{x \in \mathbf{C} : |x| > R\}$. Deduce that if $|x| > R$ then

$$y^2 = (x - \alpha_1)\ldots(x - \alpha_{2g})$$

if and only if

$$y = \pm x^g \sqrt{(1 - \frac{\alpha_1}{x})\ldots(1 - \frac{\alpha_{2g}}{x})}.$$

(ii). Show that the resolution of singularities $\pi : \tilde{C} \to C$ of C constructed in §7.1 is not a homeomorphism, but instead maps two elements of $\tilde{C}$ to the singular point $[0,1,0]$ of C (cf. figure 4.2)

What is the genus of $\tilde{C}$?

7.9. (i). Let C be the hyperelliptic curve in $\mathbf{P_2}$ defined by

$$y^2 z^2 = (x - \alpha_1 z)(x - \alpha_2 z)(x - \alpha_3 z)(x - \alpha_4 z)$$

where $\alpha_1, \dots, \alpha_4$ are distinct complex numbers and $\alpha_4 \neq 0$. Show that there is a nonsingular cubic curve D in $\mathbf{P_2}$ of the form

$$y^2 z = (x - \beta_1 z)(x - \beta_2 z)(x - \beta_3 z)$$

where

$$\beta_i = \frac{\alpha_i}{\alpha_i - \alpha_4}, \quad i = 1, 2, 3$$

and a continuous map $f : D \to C$ given by

$$f[x, y, z] = [\alpha_4 x(x - z), \gamma y z, (x - z)^2], \quad [x, y, z] \neq [0, 1, 0]$$

$$f[0, 1, 0] = [\alpha_4, 0, 1]$$

for some $\gamma \in \mathbf{C}$, such that

$$f : D - f^{-1}\{[0,1,0]\} \to C - \{[0,1,0]\}$$

is a holomorphic bijection and $f^{-1}\{[0,1,0]\}$ consists of exactly two points of D. [The existence of such a map is essentially the reason why the degrees 3 and 4 are grouped together in the definition of elliptic and hyperelliptic curves in $\mathbf{P_2}$].

(ii). Use exercise 7.8 to deduce that D is biholomorphic to the resolution of singularities $\tilde{C}$ of C constructed as in §7.1.

[N.B. Any Riemann surface which is biholomorphic to the resolution of singularities of C constructed as in §7.1 is also called a resolution of singularities of C.]

Appendix A

Algebra

Lemma A.1 *Let P and Q be polynomials in N variables. If P is homogeneous of degree d and nonzero and if Q divides P then Q is also homogeneous.*

Proof. Suppose that $P = QR$ where R is another polynomial in N variables. There are nonnegative integers g, h, m, n such that $g \leq h$ and $m \leq n$ and

$$Q = Q_g + Q_{g+1} + \ldots + Q_h$$
$$R = R_m + R_{m+1} + \ldots + R_n$$

where Q_i and R_j are homogeneous polynomials of degrees i and j if $g \leq i \leq h$ and $m \leq j \leq n$, and Q_g, Q_h, R_m, R_n are nonzero. Then

$$P = \sum_{g \leq i \leq h, m \leq j \leq n} Q_i R_j$$

and $Q_i R_j$ is a homogeneous polynomial of degree $i + j$. Equating the terms of degree e on each side we get

$$0 = \sum_{i+j=e} Q_i R_j$$

if $e \neq d$. But also we have

$$\sum_{i+j=g+m} Q_i R_j = Q_g R_m \neq 0$$

and

$$\sum_{i+j=h+n} Q_i R_j = Q_h R_n \neq 0$$

so $g + m = d = h + n$. Since $g \leq h$ and $m \leq n$ this implies that $g = h$ and $m = n$, so Q and R are both homogeneous. $\square$

If R is a commutative ring with identity then $R[x]$ denotes the ring of polynomials in x with coefficients in R. If $f(x) \in R[x]$ then $f(x)$ is called *primitive* if the only common factors in R of all the coefficients of $f(x)$ are units. (For example any monic polynomial is primitive.) $f(x)$ is called *irreducible* if it has no factors other than units in $R[x]$ and multiples of itself by units in $R[x]$.

Lemma A.2 (Gauss' lemma) *If R is a unique factorisation domain and $g(x), h(x) \in R[x]$ are primitive then so is their product $g(x)h(x)$.*

Corollary A.3 *Let R be a unique factorisation domain and K a field containing R such that every element of K can be written as ab^{-1} where $a, b \in R$ and $b \neq 0$. If $g(x) \in R[x]$ is primitive then $g(x)$ is irreducible in $R[x]$ if and only if it is irreducible in $K[x]$. Thus any $f(x) \in R[x]$ can be written as*

$$f(x) = \lambda g_1(x) \ldots g_k(x)$$

where $\lambda \in R$ and $g_1(x), \ldots, g_k(x) \in R[x]$ are primitive in $R[x]$ and irreducible in $K[x]$.

Corollary A.4 *Let R and K be as in the statement of corollary A.3. Then two polynomials $f(x)$ and $g(x)$ in $R[x]$ have a nonconstant common factor as elements of $R[x]$ if and only if they have a nonconstant common factor as elements of $K[x]$.*

Lemma A.5 *If R is a unique factorisation domain then so is $R[x]$.*

Corollary A.6 *If K is a field then the ring $K[x_1, \ldots, x_n]$ of polynomials in n variables with coefficients in K is a unique factorisation domain.*

The proofs of these results can be found in any standard algebra text (e.g. [Herstein 75], §3.11).

Appendix B

Complex analysis

Theorem B.1 (The implicit function theorem for polynomials)
(i) Let $A(z, w)$ be a polynomial with complex coefficients in two variables z and w. Suppose that

$$A(z_0, w_0) = 0 \neq \frac{\partial A}{\partial w}(z_0, w_0).$$

Then there is a holomorphic function $f : U \to V$ where U and V are open neighbourhoods of z_0 and w_0 in $\mathbb{C}$ such that

$$f(z_0) = w_0$$

and if $z \in U$ and $w \in V$ then

$$f(z) = w$$

implies

$$A(z, w) = 0.$$

(ii) Moreover

$$A(z, w) = (w - f(z))B(z, w)$$

where $B(z, w)$ is a polynomial in w whose coefficients are holomorphic functions of z.

Proof. One can deduce this theorem from the implicit function theorem for smooth functions of two real variables, but there is also a direct proof involving complex analysis.

Since

$$\frac{\partial A}{\partial w}(z_0, w_0) \neq 0$$

the polynomial $A(z_0, w)$ in w is not constant. So by the theorem of isolated zeros ([Priestley 85] 5.14) there is some $\varepsilon > 0$ such that

$$0 < \mid w - w_0 \mid \leq \varepsilon \Rightarrow A(z_0, w) \neq 0.$$

229

Thus since A is a continuous function of z and w if $|w - w_0| = \varepsilon$ there exists $\delta_w > 0$ such that

$$\max(|v - w|, |z - z_0|) \leq \delta_w \Rightarrow A(z, v) \neq 0.$$

The compact subset

$$S = \{w \in \mathbf{C} : |w - w_0| = \varepsilon\}$$

of $\mathbf{C}$ is contained in the union of the open subsets

$$\{v \in \mathbf{C} : |v - w| < \delta_w, w \in S\}$$

and hence there exists a finite subset

$$\{w_1, \ldots, w_k\}$$

of S such that

$$S \subseteq \bigcup_{1 \leq i \leq k} \{w \in \mathbf{C} : |w - w_i| < \delta_{w_i}\}.$$

Let

$$\delta = \min(\delta_{w_1}, \ldots, \delta_{w_k}) > 0.$$

Then

$$|w - w_0| = \varepsilon, \ |z - z_0| < \delta \Rightarrow A(z, w) \neq 0.$$

Define $\gamma : [0, 1] \to \mathbf{C}$ by

$$\gamma(t) = w_0 + \varepsilon e^{2\pi i t}.$$

Cauchy's residue theorem (theorem 5.4) implies that for each fixed $z \in \mathbf{C}$ satisfying $|z - z_0| < \delta$ the number of zeros of the function $A(z, w)$ of w (counted with multiplicities) inside γ is

$$n(z) = (2\pi i)^{-1} \int_\gamma \frac{\partial A}{\partial w}(z, w) A(z, w)^{-1} \, dw.$$

Since

$$A(z_0, w_0) = 0 \neq \frac{\partial A}{\partial w}(z_0, w_0)$$

the function $A(z_0, w)$ has a zero of multiplicity one at w_0. By our choice of ε it has no other zeros inside γ so

$$n(z_0) = 1.$$

Since $\gamma([0, 1])$ is compact it is easy to check that $n(z)$ is a continuous, integer-valued function of z in

$$U = \{z \in \mathbf{C} : |z - z_0| < \delta\}$$

and hence $n(z)$ is constant. Thus if $z \in U$ then $n(z) = 1$ and so there is a unique complex number, $f(z)$ say, such that

$$f(z) \in V = \{w \in \mathbf{C} \ : \mid w - w_0 \mid < \varepsilon\}$$

and $A(z, f(z)) = 0$. Moreover by Cauchy's residue theorem

$$f(z) = (2\pi i)^{-1} \int_\gamma w \frac{\partial A}{\partial w}(z, w) A(z, w)^{-1} \, dw.$$

Since $\gamma([0, 1])$ is compact we can differentiate under the integral sign with respect to z. Thus $f : U \to V$ is holomorphic and the proof of (i) is complete.

To prove (ii) we consider the ring R of functions

$$g : D \to \mathbf{C},$$

where

$$D = \{z \in \mathbf{C} \ : \mid z - z_0 \mid \leq \frac{1}{2}\delta\},$$

such that g extends to a holomorphic function on some open neighbourhood of D in $\mathbf{C}$. The units of R are the functions which are nowhere zero, and the primes are (up to multiplication by units) the functions $z - a$ with $a \in D$. It is not difficult to check that R is a unique factorisation domain. Moreover every element of the field K of functions $g : D \to \mathbf{C}$ which extend to meromorphic functions on open neighbourhoods of D in $\mathbf{C}$ can be expressed as $\frac{h}{k}$ for some $h, k \in R$. If we regard $A(z, w)$ and $w - f(z)$ as elements of the ring $R[w]$ of polynomials in w with coefficients in R then $w - f(z)$ is primitive and divides $A(z, w)$ in $K[w]$ (by the division algorithm for polynomials over a field) so it divides $A(z, w)$ in $R[w]$ by corollary A.3 in Appendix A. The result follows. $\square$

The argument used to prove theorem B.1 also gives the following result.

Theorem B.2 *Let $A(z, w)$ be a polynomial with complex coefficients in two variables z and w such that*

$$A(z_0, w_0) = 0$$

and the polynomial $A(z_0, w)$ in w has a zero of order m at w_0. Then given any $\varepsilon > 0$ there exists $\delta > 0$ such that if $\mid z - z_0 \mid < \delta$ then the polynomial $A(z, w)$ in w has at least m zeros (counting multiplicities) in the disc

$$\{w \in \mathbf{C} \ : \mid w - w_0 \mid < \varepsilon\}.$$

Remark B.3 For the proof of B.1(i) and B.2 we did not need $A(z,w)$ to be a polynomial function of z and w. It would have been enough to assume that the functions $A(z,w)$ and

$$\frac{\partial A}{\partial w}(z,w)$$

are continuous in (z,w) and that for fixed w they are holomorphic as functions of z and for fixed z they are holomorphic as functions of w.

Corollary B.4 *Let $A(z,w)$ be a polynomial with complex coefficients in z and w such that for any fixed $z \in \mathbf{C}$ the polynomial $A(z,w)$ in w is monic of degree n. Let*

$$C = \{(z,w) \in \mathbf{C}^2 \ : \ A(z,w) = 0\}$$

and define $\phi : C \to \mathbf{C}$ by

$$\phi(z,w) = z.$$

Then any $z_0 \in \mathbf{C}$ has an open neighbourhood U in $\mathbf{C}$ such that each connected component of $\phi^{-1}(U)$ contains at most one point of $\phi^{-1}(\{z_0\})$.

Proof. If

$$\phi^{-1}(\{z_0\}) = \{(z_0,w_1),\ldots,(z_0,w_k)\}$$

then

$$A(z_0,w) = \prod_{1\leq i\leq k} (w - w_i)^{m_i}$$

where $m_1,\ldots,m_k$ are positive integers such that

$$m_1 + \ldots m_k = n.$$

Choose $\varepsilon > 0$ such that $\mid w_i - w_j \mid > 2\varepsilon$ if $i \neq j$. Then by theorem B.2 there is some $\delta > 0$ such that if $\mid z - z_0 \mid < \delta$ the polynomial $A(z,w)$ in w has at least m_i roots in the disc

$$D_i = \{w \in \mathbf{C} \ : \mid w - w_i \mid < \varepsilon\}$$

when $1 \leq i \leq k$. Since the discs D_i are disjoint and the sum of the m_i is n, this means that if $\mid z - z_0 \mid < \delta$ then all the roots of $A(z,w)$ lie in

$$D_1 \cup \ldots \cup D_k,$$

and hence

$$\phi^{-1}(\{z \in \mathbf{C} \ : \mid z - z_0 \mid < \delta\}) \subseteq \mathbf{C} \times (D_1 \cup \ldots \cup D_k).$$

Therefore every connected component of

$$\phi^{-1}(\{z \in \mathbf{C} \ : \mid z - z_0 \mid < \delta\})$$

is a subset of $\mathbf{C} \times D_i$ for some $1 \leq i \leq k$, and hence contains at most one point of $\phi^{-1}(\{z_0\})$. $\square$

Theorem B.5 (Inverse function theorem)
(i) Let $f : U \to V$ be a holomorphic bijection between open subsets U and V of $\mathbf{C}$. Then

$$f'(z) \neq 0$$

for all $z \in U$ and the inverse

$$f^{-1} : V \to U$$

of f is holomorphic.
(ii) Let $f : U \to \mathbf{C}$ be a holomorphic function, defined on a neighbourhood U of a in $\mathbf{C}$, such that

$$f'(a) \neq 0.$$

Then the restriction of f to any sufficiently small open neighbourhood of a in U is a holomorphic bijection onto an open neighbourhood of $f(a)$ in $\mathbf{C}$.

Proof. See for example [Priestley 85] 10.6 (p.168), exercise 10.8 (p.192) and 10.25 (p.182). □

Theorem B.6 *Suppose that U is an open subset of $\mathbf{C}$ and that $f : U \to \mathbf{C}$ is continuous. If the restriction of f to*

$$U - \{a_1, \ldots, a_m\}$$

is holomorphic for some $a_1, \ldots, a_m$ in U then f is holomorphic.

Proof. Without loss of generality we may assume that $m = 1$ and that

$$U = \{z \in \mathbf{C} \ : \ |z - a| < r\}$$

for some $r > 0$ where $a = a_1$. By Laurent's theorem we can express $f(z)$ as a power series

$$f(z) = \sum_{k \in \mathbf{Z}} c_k (z - a)^k$$

when $0 < |z - a| < r$, where

$$c_k = (2\pi)^{-1} \int_0^{2\pi} f(a + \varepsilon e^{it})(\varepsilon e^{it})^{-k} \, dt$$

for any $0 < \varepsilon < r$. Since f is continuous and

$$\{z \in \mathbf{C} \ : \ |z - a| \leq \tfrac{1}{2}r\}$$

is a compact subset of U there is some M such that

$$|z - a| \leq \tfrac{1}{2}r \Rightarrow |f(z)| \leq M.$$

Then if $0 < \varepsilon < \frac{1}{2}r$ we have

$$
\begin{aligned}
| \, c_k \, | &\leq (2\pi)^{-1} \int_0^{2\pi} | \, f(a + \varepsilon e^{it}) \, | \, \varepsilon^{-k} \, dt \\
&\leq (2\pi)^{-1} \int_0^{2\pi} M \varepsilon^{-k} \, dt \\
&= M \varepsilon^{-k}
\end{aligned}
$$

and this tends to 0 with ε if $k < 0$. Hence $c_k = 0$ if $k < 0$, so

$$
f(z) = \sum_{k \geq 0} c_k (z - a)^k
$$

is holomorphic on U.

Appendix C

Topology

C.1 Covering projections

Definition C.1 (cf. [Spanier 66] §2.1) *A continuous map* $\pi : Y \to X$ *between topological spaces* X *and* Y *is called a* covering projection *if each* $x \in X$ *has an open neighbourhood* U *in* X *such that* $\pi^{-1}(U)$ *is a disjoint union of open subsets of* Y *each of which is mapped homeomorphically onto* U *by* π.

Examples C.2 (1) The map $\pi : \mathbf{C} \to \mathbf{C}/\Lambda$ defined by $\pi(z) = \Lambda + z$ is a covering projection for any lattice Λ in $\mathbf{C}$ (see 5.21 and 5.42).
(2) Let C be a nonsingular projective curve in $\mathbf{P}_2$ not containing $[1, 0, 0]$ and define $\phi : C \to \mathbf{P}_1$ by

$$\phi[x, y, z] = [x, z]$$

as in §4.2. If R is the set of ramification points of ϕ then

$$\phi : C - R \to \mathbf{P}_1 - \phi(R)$$

is a covering projection. (This can be proved by a modification of the proof of 5.28.)

Lemma C.3 *Let* $\pi : Y \to X$ *be a covering projection and let* $f : [0, 1] \to X$ *be continuous. Given any* $y \in Y$ *such that* $\pi(y) = f(0)$ *there is a unique continuous map* $F : [0, 1] \to Y$ *such that* $F(0) = y$ *and*

$$\pi \circ F = f.$$

Proof. Since π is a covering projection X has an open cover $\mathcal{F}$ such that if $U \in \mathcal{F}$ then $\pi^{-1}(U)$ is a disjoint union of open subsets of Y each mapped homeomorphically onto U by π. Since $[0, 1]$ is compact there exist $0 = t_0 < t_1 < \ldots < t_k = 1$ and $U_1, \ldots, U_k$ in $\mathcal{F}$ such that

$$[t_{i-1}, t_i] \subseteq f^{-1}(U_i)$$

if $1 \leq i \leq k$. Then since $\pi(y) = f(0)$ we have $y \in \pi^{-1}(U_i)$ which is the disjoint union of open subsets V_j, say, for j in some indexing set J, such that $\pi : V_j \to U_1$ is a homeomorphism with inverse $\phi_j : U_1 \to V_j$ for each $j \in J$. Let j_0 be the element of J such that $y \in V_{j_0}$. Then the composition

$$F_1 = \phi_{j_0} \circ f \mid_{[0,t_1]} : [0, t_1] \to V_1 \subseteq Y$$

satisfies $F_1(0) = y$ and $\pi \circ F_1$ is the restriction of f to $[0, t_1]$. By induction we can find continuous maps

$$F_i : [t_{i-1}, t_i] \to Y$$

satisfying $F_i(t_{i-1}) = F_{i-1}(t_{i-1})$ and such that $\pi \circ F_i$ is the restriction of f to $[t_{i-1}, t_i]$. Since a map from the interval $[0, 1]$ to Y is continuous if and only if its restriction to each subinterval $[t_{i-1}, t_i]$ is continuous the map $F : [0, 1] \to Y$ defined by

$$F(t) = F_i(t) \quad \text{if } t \in [t_{i-1}, t_i]$$

is continuous and satisfies $F(0) = y$ and $\pi \circ F = f$.

Now suppose that $G : [0, 1] \to Y$ is another continuous map such that $\pi \circ G = f$. Let

$$R = \{t \in [0, 1] \ : \ F(t) = G(t)\}.$$

We have

$$G([0, t_1]) \subseteq \pi^{-1}(f[0, t_1]) \subseteq \pi^{-1}(U_1)$$

which is a disjoint union of the open subsets V_j of Y. Since $G([0, t_1])$ is connected this implies that

$$G([0, t_1]) \subseteq V_{j_1}$$

for some $j_1 \in J$, and since $\pi : V_{j_1} \to U_1$ is a homeomorphism with inverse ϕ_{j_1} we must have

$$G(t) = \phi_{j_1} \circ f(t)$$

for all $t \in [0, 1]$. Then either $j_1 = j_0$, in which case

$$[0, t_1] \subseteq R$$

or else $j_1 \neq j_0$, in which case V_{j_1} and V_{j_0} are disjoint and hence

$$[0, t_1] \subseteq [0, 1] - R.$$

The same is true for each interval $[t_{i-1}, t_i]$. This implies that R is either the whole interval $[0, 1]$ or R is empty. If $G(0) = F(0)$ then R is not empty so we must have $G = F$. This proves the uniqueness of the function F. $\square$

Definition C.4 *Let*

$$\Delta = \{(x,y) \in \mathbf{R}^2 \ : \ x \geq 0, y \geq 0, x + y \leq 1\},$$

$$\Delta^0 = \{(x,y) \in \mathbf{R}^2 \ : \ x > 0, y > 0, x + y < 1\}$$

and

$$\partial\Delta = \Delta - \Delta^0.$$

A topological space X is called simply connected *if any continuous map $g : \partial\Delta \to X$ can be extended to a continuous map*

$$g : \Delta \to X.$$

(Equivalently any loop in X can be continuously deformed to a point in X.)

Examples C.5 Any interval in $\mathbf{R}$ is simply connected. Δ is simply connected. More generally any convex subset A of $\mathbf{C}$ is simply connected. So is the Riemann sphere $\mathbf{P}_1 = \mathbf{C} \cup \{\infty\}$.

Lemma C.6 *(i) Let $\pi : Y \to X$ be a covering projection and let $f : A \to X$ be a continuous map. Suppose that A is simply connected, path connected and locally path connected (i.e. every $a \in A$ has arbitrarily small path connected open neighbourhoods in A). Then given any $a \in A$ and $y \in Y$ such that $f(a) = \pi(y)$ there is a unique continuous map $F : A \to Y$ such that $F(a) = y$ and*

$$\pi \circ F = f.$$

(ii) If moreover f is a homeomorphism onto its image $f(A)$, then F is a homeomorphism onto a connected component of $\pi^{-1}(f(A))$.

Proof. (i) Since A is path connected, if $b \in A$ then there is a continuous map $g : [0,1] \to A$ such that $g(0) = a$ and $g(1) = b$. By lemma C.3 there is a unique continuous map $G : [0,1] \to Y$ such that $\pi \circ G = f \circ g$ and $G(0) = y$. We shall show that $G(1)$ does not depend on the choice of path g from a to b, but only on b itself.

So suppose that $h : [0,1] \to A$ is another map from a to b, and let $H : [0,1] \to Y$ be the unique continuous map such that $H(0) = y$ and $\pi \circ H = f \circ h$. Then there is a continuous map

$$e : \partial\Delta \to A$$

defined by

$$e(t,0) = g(t), \ \ e(0,t) = h(t), \ \ e(t,1-t) = b$$

if $t \in [0,1]$. Since A is simply connected e extends to a continuous map $e : \Delta \to A$. By modifying the proof of lemma C.3 we find that there is a unique continuous map $E : \Delta \to Y$ such that $\pi \circ E = f \circ e$ and $E(0,0) = a$, and

$$E(t,0) = G(t), \quad E(0,t) = H(t)$$

for all $t \in [0,1]$. Thus there is a continuous map $k : [0,1] \to Y$ defined by

$$k(t) = E(1 - t, t)$$

such that $k(0) = G(1)$ and

$$\pi \circ k(t) = \pi \circ E(1 - t, t) = f \circ e(1 - t, t) = f(a)$$

for $t \in [0,1]$. By the uniqueness part of lemma C.3 it follows that $k(t) = G(1)$ for all $t \in [0,1]$, and hence that

$$H(1) = E(0,1) = k(1) = G(1).$$

This shows that there is a well-defined map $F : A \to Y$ given by $F(b) = G(1)$ where G is chosen as above. By its construction F satisfies $F(a) = y$ and $\pi \circ F = f$, and uniqueness follows from lemma C.3. Thus it remains to show that F is continuous at any $b \in A$.

Let W be any open neighbourhood of $F(b)$ in Y. Since π is a covering projection there is an open neighbourhood V of $F(b)$ in W such that V is mapped homeomorphically by π onto an open neighbourhood U of $\pi(F(b)) = f(b)$ in X. Let $\phi : U \to V$ be the inverse of $\pi : V \to U$. Then $f^{-1}(U)$ is an open neighbourhood of b in A, so it contains a path connected open neighbourhood T of b in A because A is locally path connected. If $c \in T$ then there is a path γ_c from b to c in T, and the join $g.\gamma_c$ of this path γ_c with the path g from a to b in A is a path in A from a to c satisfying

$$f \circ (g.\gamma_c) = \pi \circ (G.(\phi \circ f \circ \gamma_c)(1)).$$

Hence $F(c) = \phi \circ f \circ \gamma(1)$ which is an element of V and hence of W. Thus $F^{-1}(W)$ contains the open neighbourhood T of b in X. This shows that F is continuous at b and completes the proof of (i).

(ii) Now suppose that $f : A \to X$ is a homeomorphism onto its image. Since the restriction of the covering projection π to $\pi^{-1}(f(A))$ is a covering projection we may assume that $f(A) = X$. Note that since A is connected so are $X = f(A)$ and $F(A)$.

Any $x \in X$ has a connected open neighbourhood U_x in X such that $\pi^{-1}(U_x)$ is a disjoint union of open subsets of Y each of which is mapped homeomorphically by π onto U_x. Since $F(f^{-1}(U_x))$ is connected and is contained in $\pi^{-1}(U_x)$ it is contained in one of these open subsets, V_x say. Since $\pi \circ F = f$ and $\pi : V_x \to U_x$ is a homeomorphism it follows that $F : f^{-1}(U_x) \to V_x$ is the composition of f with the inverse of $\pi : V_x \to U_x$ and hence

$$F(A) \cap \pi^{-1}(U_x) = F(f^{-1}(U_x)) = V_x$$

is both open and closed in $\pi^{-1}(U_x)$. It follows that $F(A)$ is both open and closed in Y, so it must be a connected component of Y.

F is injective because $\pi \circ F$ is f which is a homeomorphism onto its image in X. Therefore $F : A \rightarrow F(A)$ is a continuous bijection whose inverse is the continuous map $\pi : F(A) \rightarrow A$. Thus F is a homeomorphism onto its image. $\square$

Lemma C.7 *Let $\pi : Y \rightarrow X$ be a continuous map and suppose that every $x \in X$ has an open neighbourhood U in X such that each connected component of $\pi^{-1}(U)$ contains at most one point of $\pi^{-1}(x)$. Suppose that Y is compact and that V is an open subset of X such that $\pi : \pi^{-1}(V) \rightarrow V$ is a covering projection. If $f : [0,1] \rightarrow X$ is continuous and $f^{-1}(V)$ contains the open interval $(0,1)$ then given $\tau \in (0,1)$ and $y \in Y$ such that $\pi(y) = f(\tau)$ there is a unique continuous map $F : [0,1] \rightarrow Y$ such that $F(\tau) = y$ and $\pi \circ F = f$.*

Proof. By lemma C.6 there is a unique continuous map $F : (0,1) \rightarrow Y$ such that $F(\tau) = y$ and

$$\pi \circ F = f \mid_{(0,1)} .$$

It suffices to show that $F(t)$ tends to a unique limit p in Y as $t \rightarrow 0$ and a unique limit q in Y as $t \rightarrow 1$, so that F can be extended uniquely to a continuous map $F : [0,1] \rightarrow Y$ by setting $F(0) = p$ and $F(1) = q$.

By hypothesis $f(0)$ has an open neighbourhood U in X such that each connected component of $\pi^{-1}(U)$ contains at most one point of $\pi^{-1}(f(0))$. If $\delta > 0$ is sufficiently small then

$$f((0,\delta]) \subseteq U$$

and so

$$F((0,\delta]) \subseteq \pi^{-1}(U).$$

Since $F((0,\delta])$ is connected it is contained in a component, W say, of $\pi^{-1}(U)$.

Let $t_1, t_2, \ldots$ be a sequence in $(0,\delta]$ which tends to 0. Then since Y is compact there is a subsequence

$$t_{n_1}, t_{n_2}, \ldots$$

such that

$$F(t_{n_k})$$

converges to some $p \in Y$ as k tends to infinity. Then

$$\pi(p) = \lim \pi \circ F(t_{n_k}) = \lim f(t_{n_k}) = f(0)$$

so $p \in \pi^{-1}(U)$. Since $F(t_n) \in W$ for all n and W is an open and closed subset of $\pi^{-1}(U)$ it follows that p must be the *unique* point of $\pi^{-1}(f(0))$ lying in W. This implies that $F(t) \rightarrow p$ as $t \in [0,\delta)$ tends to 0. A similar argument deals with the limit as $t \rightarrow 1$, and hence the proof is complete. $\square$

Remark C.8 It is easy to modify this argument to apply when $[0,1]$ is replaced by Δ and $(0,1)$ is replaced by

$$\Delta - \{(0,0),(1,0),(0,1)\}.$$

Remark C.9 By corollary B.4 of Appendix B the hypotheses of this lemma are satisfied by the map $\phi : C \to \mathbf{P}_1$ defined as in example C.2(2).

C.2 The genus is a topological invariant

Let X be a compact Riemann surface. As in §4.3 we say that a triangulation T of X is given by the following data:

(a) a finite nonempty set V of points of X called *vertices*,
(b) a finite nonempty set E of continuous maps $e : [0,1] \to X$ called *edges*,
(c) a finite nonempty set F of continuous maps $f : \Delta \to X$ where

$$\Delta = \{(x,y) \in \mathbf{R}^2 \ : \ x \geq 0, y \geq 0, x + y \leq 1\}$$

called *faces*,

   satisfying

(i) $V = \{e(0) : e \in E\} \cup \{e(1) : e \in E\}$;
(ii) if $e \in E$ then $e(t) \in V$ if and only if $t \in \{0,1\}$, and the restriction of e to $(0,1)$ is a homeomorphism onto its image in X;
(iii) if $e, \tilde{e}$ are distinct edges then $e(t) \neq \tilde{e}(s)$ for all $s, t \in (0,1)$;
(iv) if $f : \Delta \to X$ is a face then the restriction of f to

$$\Delta^0 = \{(x,y) \in \mathbf{R}^2 \ : \ x > 0, y > 0, x + y < 1\}$$

is a homeomorphism of Δ^0 onto a connected component K_f of

$$X - \bigcup_{e \in E} e([0,1]),$$

and if $\sigma_i : [0,1] \to \Delta$ for $1 \leq i \leq 3$ and $r : [0,1] \to [0,1]$ are defined by

$$\sigma_1(t) = (t,0) \quad \sigma_2(t) = (1-t,t) \quad \sigma_3(t) = (0,1-t), \quad r(t) = 1 - t$$

then either $f \circ \sigma_i$ or $f \circ \sigma_i \circ r$ is an edge $e_f^i \in E$ for $1 \leq i \leq 3$;
(v) the mapping $f \mapsto K_f$ from F to the set of connected components of $X - \bigcup_{e \in E} e([0,1])$ is a bijection;
(vi) for every $e \in E$ there is exactly one face $f_e^+ \in F$ such that $e = f_e^+ \circ \sigma_i$ for some $i \in \{1,2,3\}$ and exactly one face $f_e^- \in F$ such that $e = f_e^- \circ \sigma_i \circ r$ for some $i \in \{1,2,3\}$.

The Euler number $\chi_T(X)$ of X with respect to the triangulation T is

$$\chi_T(X) = \#V - \#E + \#F.$$

The aim of this section of the appendix is to prove

Theorem C.10 $\chi_T(X)$ *depends only on* X, *not on the choice of triangulation* T.

In order to prove this theorem we introduce complex vector spaces

$$C_0^T(X), C_1^T(X), C_2^T(X)$$

with bases V, E and F respectively. Thus an element of $C_0^T(X)$ can be written uniquely as a linear combination

$$\sum_{v \in V} \lambda_v v$$

with $\lambda_v \in \mathbf{C}$ for each $v \in V$, and similarly

$$C_1^T(X) = \{\sum_{e \in E} \lambda_e e \; : \; \lambda_e \in \mathbf{C} \;\; \forall e \in E\}$$

$$C_2^T(X) = \{\sum_{f \in F} \lambda_f f \; : \; \lambda_f \in \mathbf{C} \;\; \forall f \in F\}.$$

There are linear maps $\partial_2^T : C_2^T(X) \to C_1^T(X), \partial_1^T : C_1^T(X) \to C_0^T(X)$ and $\partial_0^T : C_0^T(X) \to \mathbf{C}$ defined as follows.

$$\begin{aligned}
\partial_0^T(\sum_{v \in V} \lambda_v v) &= \sum_{v \in V} \lambda_v \\
\partial_1^T(\sum_{e \in E} \lambda_e e) &= \sum_{e \in E} \lambda_e(e(1) - e(0)) \\
\partial_2^T(\sum_{f \in F} \lambda_f f) &= \sum_{f \in F} \lambda_f(\pm e_f^1 \pm e_f^2 \pm e_f^3)
\end{aligned}$$

where the sign $\pm$ in front of e_f^i is positive if $f \circ \sigma_i = e_f^i$ and is negative if $f \circ \sigma_i \circ r = e_f^i$. Note that

$$\partial_0^T \partial_1^T(e) = \partial_0^T(e(1) - e(0)) = 0$$

for all $e \in E$ and

$$\begin{aligned}
\partial_1^T \partial_2^T(f) &= \partial_1^T(\pm e_f^1 \pm e_f^2 \pm e_f^3) \\
&= f(1,0) - f(0,0) + f(0,1) - f(1,0) + f(0,0) - f(0,1) \\
&= 0
\end{aligned}$$

for all $f \in F$. Thus

$$\operatorname{im} \partial_1^T \subseteq \ker \partial_0^T, \;\; \operatorname{im} \partial_2^T \subseteq \ker \partial_1^T.$$

Lemma C.11

$$\chi_T(X) = \dim\left(\frac{\ker \partial_0^T}{\operatorname{im} \partial_1^T}\right) - \dim\left(\frac{\ker \partial_1^T}{\operatorname{im} \partial_2^T}\right) + k + 1$$

where k is the number of connected components of X.

Proof. By the rank-nullity formula

$$\#V = \dim C_0^T(X) = \dim\ker \partial_0^T + \dim\operatorname{im} \partial_0^T$$

$$\#E = \dim C_1^T(X) = \dim\ker \partial_1^T + \dim\operatorname{im} \partial_1^T$$

$$\#F = \dim C_2^T(X) = \dim\ker \partial_2^T + \dim\operatorname{im} \partial_2^T$$

so

$$\begin{aligned}
\chi_T(X) &= \#V - \#E + \#F \\
&= \dim\left(\tfrac{\ker \partial_0^T}{\operatorname{im} \partial_1^T}\right) - \dim\left(\tfrac{\ker \partial_1^T}{\operatorname{im} \partial_2^T}\right) + \dim\operatorname{im} \partial_0^T + \dim\ker \partial_2^T.
\end{aligned}$$

Since $V \neq \emptyset$ the map $\partial_0^T : C_0^T(X) \to \mathbf{C}$ is surjective so

$$\dim\operatorname{im} \partial_0^T = 1.$$

Moreover by (vi) in the definition of a triangulation we have

$$\partial_2^T\Big(\sum_{f \in F} \lambda_f f\Big) = \sum_{e \in E}(\lambda_{f_e^+} - \lambda_{f_e^-})e$$

which vanishes if and only if

$$\lambda_{f_e^+} = \lambda_{f_e^-} \quad \forall e \in E.$$

Let $X_1,\ldots,X_k$ be the connected components of X. If $f : \Delta \to X$ is a face then $f(\Delta)$ is a closed connected subset of X so $f(\Delta)$ is contained in one of the components X_i of X. Moreover

$$X = \bigcup_{f \in F} f(\Delta)$$

and if $f \neq \tilde{f}$ then $f(\Delta) \cap \tilde{f}(\Delta) \neq \emptyset$ if and only if there is some $e \in E$ such that

$$\{f_e^-, f_e^+\} = \{f, \tilde{f}\}.$$

It follows that $f(\Delta)$ and $\tilde{f}(\Delta)$ are contained in the same connected component of X if and only if there is a sequence

$$f = f_0, f_1, f_2, \ldots f_l = \tilde{f}$$

in F and $e_1, \ldots, e_l$ in E such that

$$\{f_{e_j}^-, f_{e_j}^+\} = \{f_j, f_{j+1}\}$$

for $1 \leq j \leq l$. Therefore an element $\sum_{f \in F} \lambda_f f$ of $C_2^T(X)$ satisfies

$$\lambda_{f_e^+} = \lambda_{f_e^-} \quad \forall e \in E$$

if and only if there exist $\mu_1, \ldots, \mu_k \in \mathbb{C}$ such that $\lambda_f = \mu_i$ when $f(\Delta) \subseteq X_i$, i.e. if and only if

$$\sum_{f \in F} \lambda_f f = \sum_{1 \leq i \leq k} \mu_i \Big(\sum_{f(\Delta) \subseteq X_i} f \Big).$$

Thus $\dim \ker \partial_2^T = k$. The result follows. $\square$

The idea now is to identify the vector spaces

$$\frac{\ker \partial_0^T}{\operatorname{im} \partial_1^T}, \quad \frac{\ker \partial_1^T}{\operatorname{im} \partial_2^T}$$

with vector spaces defined similarly but which are independent of the triangulation T. To this end let $C_0(X)$ be a complex vector space with basis X. This means that any element of $C_0(X)$ can be uniquely expressed in the form

$$\sum_{x \in X} \lambda_x x$$

where $\lambda_x \in \mathbb{C}$ for each $x \in X$ and $\lambda_x = 0$ for all but finitely many $x \in X$. Conversely every such expression defines an element of $C_0(X)$. Of course $C_0(X)$ is an infinite-dimensional vector space. Similarly let $C_1(X)$ be a complex vector space with basis the set

$$\operatorname{Map}([0,1], X) = \{\sigma : [0,1] \to X : \sigma \text{ continuous}\}$$

of all continuous maps from $[0,1]$ to X, and let $C_2(X)$ be a complex vector space with basis the set

$$\operatorname{Map}(\Delta, X) = \{\Sigma : \Delta \to X : \Sigma \text{ continuous}\}$$

of all continuous maps from Δ to X. Since

$$V \subseteq X, \quad E \subseteq \operatorname{Map}([0,1], X) \text{ and } F \subseteq \operatorname{Map}(\Delta, X)$$

we can identify $C_0^T(X), C_1^T(X)$ and $C_2^T(X)$ with the subspaces of $C_0(X)$, $C_1(X)$ and $C_2(X)$ spanned by V, E and F respectively.

Since X is a basis for $C_0(X)$ there is a unique linear map $\partial_0 : C_0(X) \to \mathbb{C}$ such that

$$\partial_0(x) = 1 \quad \forall x \in X,$$

given by

$$\partial_0(\sum_{x \in X} \lambda_x x) = \sum_{x \in X} \lambda_x.$$

This sum makes sense because only finitely many of the coefficients λ_x are nonzero. Similarly since Map $([0,1], X)$ is a basis for $C_1(X)$ there is a unique linear map $\partial_1 : C_1(X) \to C_0(X)$ satisfying

$$\partial_1(\sigma) = \sigma(1) - \sigma(0)$$

for every continuous map $\sigma : [0,1] \to X$. Finally since Map $(\triangle, X)$ is a basis for $C_2(X)$ there is a unique linear map $\partial_2 : C_2(X) \to C_1(X)$ satisfying

$$\partial_2(\Sigma) = \Sigma \circ \sigma_1 + \Sigma \circ \sigma_2 + \Sigma \circ \sigma_3$$

if $\Sigma : \triangle \to X$ is continuous. Here $\sigma_i : [0,1] \to \triangle$ for $1 \le i \le 3$ are defined as before by

$$\sigma_1(t) = (t,0), \quad \sigma_2(t) = (1-t,t), \quad \sigma_3(t) = (0,1-t).$$

If $\sigma \in$ Map $([0,1], X)$ and $\Sigma \in$ Map $(\triangle, X)$ then

$$\begin{aligned}
\partial_0\partial_1(\sigma) &= \partial_0(\sigma(1) - \sigma(0)) = 1 - 1 = 0\\
\partial_1\partial_2(\Sigma) &= \partial_1(\Sigma \circ \sigma_1 + \Sigma \circ \sigma_2 + \Sigma \circ \sigma_3)\\
&= \Sigma(1,0) - \Sigma(0,0) + \Sigma(0,1) - \Sigma(0,1) + \Sigma(0,0) - \Sigma(0,1)\\
&= 0
\end{aligned}$$

so im $\partial_1 \subseteq$ ker ∂_0 and im $\partial_1 \subseteq$ ker ∂_2. Since these spaces are independent of the triangulation T, in order to prove that $\chi_T(X)$ is independent of T it suffices to prove the following result.

Proposition C.12

$$\frac{\ker \partial_1}{\text{im } \partial_2} \cong \frac{\ker \partial_1^T}{\text{im } \partial_2^T} \quad \text{and} \quad \frac{\ker \partial_0}{\text{im } \partial_1} \cong \frac{\ker \partial_0^T}{\text{im } \partial_1^T}.$$

For the proof we need three lemmas.

Lemma C.13 *If $\xi \in C_0(X)$ then there exists $\eta \in C_1(X)$ such that $\xi - \partial_1\eta \in C_0^T(X)$.*

Lemma C.14 *If $\xi \in C_1(X)$ and $\partial_1\xi \in C_0^T(X)$ then there exists $\eta \in C_2(X)$ such that $\xi - \partial_2\eta \in C_1^T(X)$.*

Lemma C.15 *If $\xi \in C_2(X)$ and $\partial_2\xi \in C_1^T(X)$ then there exists $\eta \in C_2^T(X)$ such that $\partial_2\xi = \partial_2^T\eta$.*

Proof of proposition C.12, given lemmas C.13,C.14,C.15. It suffices to show that for $j = 0, 1$,

$$\ker \partial_j = \operatorname{im} \partial_{j+1} + \ker \partial_j^T$$

and

$$\operatorname{im} \partial_{j+1} \cap \ker \partial_j^T = \operatorname{im} \partial_{j+1}^T.$$

Then the inclusions $\ker \partial_j^T \subseteq \ker \partial_j, \operatorname{im} \partial_{j+1}^T \subseteq \operatorname{im} \partial_{j+1}$ will induce isomorphisms

$$\frac{\ker \partial_j^T}{\operatorname{im} \partial_{j+1}^T} \cong \frac{\ker \partial_j}{\operatorname{im} \partial_{j+1}}.$$

Since ∂_j^T is the restriction of ∂_j to $C_j^T(X)$ for $j = 0, 1$ and since

$$\partial_j \circ \partial_{j+1} = 0, \quad \partial_j^T \circ \partial_{j+1}^T = 0,$$

it is clear that

$$\ker \partial_j \supseteq \operatorname{im} \partial_{j+1} + \ker \partial_j^T$$

$$\operatorname{im} \partial_{j+1} \cap \ker \partial_j^T \supseteq \operatorname{im} \partial_{j+1}^T.$$

Now suppose $\xi \in \ker \partial_j$. Then by lemma C.13 (when $j = 0$) and lemma C.14 (when $j = 1$) there exists $\eta \in C_{j+1}(X)$ such that

$$\xi - \partial_{j+1}\eta = \zeta \in C_j^T(X).$$

Then

$$\partial_j^T \zeta = \partial_j \zeta = \partial_j \xi - \partial_j \partial_{j+1}\eta = 0 - 0 = 0$$

so $\zeta \in \ker \partial_j^T$ and

$$\xi = \zeta + \partial_{j+1}\eta \in \ker \partial_j^T + \operatorname{im} \partial_{j+1}.$$

Next suppose $\zeta \in \ker \partial_0^T$ and $\zeta = \partial_1 \xi$ for some $\xi \in C_1(X)$. By lemma C.14 there exists $\eta \in C_2(X)$ such that $\xi - \partial_2 \eta = \chi \in C_1^T(X)$. Then

$$\zeta = \partial_1 \xi = \partial_1 (\chi + \partial_2 \eta) = \partial_1 \chi = \partial_1^T \chi.$$

Thus

$$\operatorname{im} \partial_1 \cap \ker \partial_0^T \subseteq \operatorname{im} \partial_1^T.$$

Finally suppose $\zeta \in \ker \partial_1^T$ and $\zeta = \partial_2 \xi$ where $\xi \in C_2(X)$. Then by lemma C.15 there exists $\eta \in C_2^T(X)$ such that

$$\partial_2^T \eta = \partial_2 \xi = \zeta$$

so

$$\operatorname{im} \partial_2 \cap \ker \partial_1^T \subseteq \operatorname{im} \partial_2^T.$$

This completes the proof of the proposition, given lemmas C.13,C.14,C.15. $\square$

Proof of lemma C.13. Let $X_1, \ldots, X_k$ be the connected components of X. Every component X_i must contain at least one vertex of the triangulation T: let v_i be one such vertex. Each X_i is a Riemann surface and hence is locally path-connected. Since X_i is connected it must be path-connected. Thus if $x \in X_i$ there is a continuous map $\sigma_x : [0,1] \to X_i$ such that $\sigma_x(0) = v_i$ and $\sigma_x(1) = x$. Therefore if

$$\xi = \sum_{x \in X} \lambda_x x \in C_0(X)$$

then

$$\eta = \sum_{x \in X} \lambda_x \sigma_x \in C_1(X)$$

and

$$\begin{aligned}\xi - \partial_1 \eta &= \sum_{x \in X} \lambda_x x - \sum_{1 \le i \le k} \sum_{x \in X_i} \lambda_x (x - v_i) \\ &= \sum_{1 \le i \le k} \left(\sum_{x \in X_i} \lambda_x \right) v_i \in C_0^T(X).\end{aligned}$$

Proof of lemma C.14. Define homeomorphisms $h_j : \Delta \to \Delta$ for $1 \le j \le 6$ by

$$\begin{array}{lll} h_1(x,y) = (x,y) & h_2(x,y) = (x, 1-x-y) & h_3(x,y) = (1-x-y, y) \\ h_4(x,y) = (y,x) & h_5(x,y) = (1-x-y, x) & h_6(x,y) = (y, 1-x-y). \end{array}$$

Let

$$\Delta^* = \{(x,y) \in \Delta : x + y < \frac{1}{2}\}$$

and if $v \in V$ define

$$\text{Star}\,(v) = \bigcup f \circ h_j(\Delta^*)$$

where the union is over all $f \in F$ and $1 \le j \le 6$ such that $f \circ h_j(0,0) = v$. Then $\text{Star}\,(v)$ is an open neighbourhood of v in X and $\text{Star}\,(v) \cap \text{Star}\,(\tilde{v}) = \emptyset$ if $v \ne \tilde{v}$. If $e \in E$ let

$$\text{Star}\,(e) = e([0,1]) \cup f_e^+(\Delta^0) \cup f_e^-(\Delta^0) \cup \text{Star}\,(e(0)) \cup \text{Star}\,(e(1))$$

(see figure C.1). Then it is easy to check that

$$\{\text{Star}\,(e) : e \in E\}$$

is an open cover of X. Thus if $\sigma : [0,1] \to X$ is continuous there exist $0 = t_0 < t_1 < \ldots < t_k = 1$ and $e_1, \ldots, e_k \in E$ such that

$$[t_{i-1}, t_i] \subseteq \sigma^{-1}(\text{Star}\,(e_i)) \quad 1 \le i \le k.$$

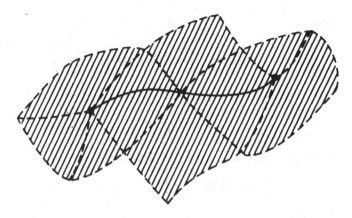

Figure C.1: Star(e)

Since

$$\sigma(t_i) \in \text{Star}\,(e_i) \cap \text{Star}\,(e_{i+1})$$

we must have either

$$\{e_i(0), e_i(1)\} \cap \{e_{i+1}(0), e_{i+1}(1)\} \neq \emptyset$$

or there exists $e \in E$ such that $\sigma(t_i) \in \text{Star}\,(e)$ and

$$\{e_i(0), e_i(1)\} \cap \{e(0), e(1)\} \neq \emptyset \neq \{e(0), e(1)\} \cap \{e_{i+1}(0), e_{i+1}(1)\}.$$

Thus by adding extra edges to the list $e_1, \ldots, e_k$ if necessary and by adding the composition of edges with the map $r : [0,1] \rightarrow [0,1]$ given by

$$r(t) = 1 - t$$

we may assume that there exist $0 = t_0 < t_1 < \ldots < t_k = 1$ and $e_1, \ldots, e_k \in E$ and $\Sigma_1, \ldots, \Sigma_k$ such that $\Sigma_i = e_i$ or $\Sigma_i = e_i \circ r$ and

$$[t_{i-1}, t_i] \subseteq \sigma^{-1}(\text{Star}\,(e_i))$$

and

$$\Sigma_{i-1}(1) = \Sigma_i(0)$$

for $1 \leq i \leq k$. Then there is a well-defined continuous map $G_\sigma : [0,1] \times [0,1]$ given as follows. If $t \in [t_{i-1}, t_i]$ and

$$\sigma(t) = f_{e_i}^{\pm} h_j(a, b)$$

for some $1 \leq j \leq 6$ such that $f_{e_i}^{\pm} h_j(1,0) = \Sigma_i(0)$ and $f_{e_i}^{\pm} h_j(0,1) = \Sigma_i(1)$ and some $(a, b) \in \Delta$ such that $a > 0$ and $b > 0$ then

$$G_\sigma(s,t) = f_{e_i}^{\pm} h_j \left(\frac{2sa}{a+b} + (1-2s)a, \frac{2sb}{a+b} + (1-2s)b \right)$$

if $s \in [0, \frac{1}{2}]$ and

$$G_\sigma(s,t) = \Sigma_i \left(\frac{(2-2s)b}{a+b} + (2s-1)\frac{(t-t_{i-1})}{(t_i-t_{i-1})} \right)$$

if $s \in [\frac{1}{2}, 1]$. If $t \in [t_{i-1}, t_i]$ and

$$\sigma(t) = f h_j(a,b)$$

for some $f \in F - \{f_{e_i}^+, f_{e_i}^-\}$ and $1 \le j \le 6$ such that

$$f h_j(0,0) \in \{\Sigma_i(0), \Sigma_i(1)\}$$

and $(a,b) \in \Delta^*$ then

$$G_\sigma(s,t) = f h_j((1-2s)a, (1-2s)b)$$

if $s \in [0, \frac{1}{2}]$ and

$$G_\sigma(s,t) = \begin{cases} \Sigma_i \left((2s-1)\frac{(t-t_{i-1})}{(t_i-t_{i-1})} \right) & \text{if } f h_j(0,0) = \Sigma_i(0) \\ \Sigma_i \left(2 - 2s + (2s-1)\frac{(t-t_{i-1})}{(t_i-t_{i-1})} \right) & \text{if } f h_j(0,0) = \Sigma_i(1). \end{cases}$$

Then we have

$$\begin{aligned} G_\sigma(0,t) &= \sigma(t) \\ G_\sigma(1,t) &= \Sigma_i \left(\frac{t-t_{i-1}}{t_i-t_{i-1}} \right) \quad \text{if } t \in [t_{i-1}, t_i]. \end{aligned}$$

If we divide the square $[0,1] \times [0,1]$ into two copies of Δ and restrict G_σ to each of these we get $\Sigma_\sigma^1, \Sigma_\sigma^2 \in C_2(X)$ such that

$$\partial_2 \Sigma_\sigma^1 + \partial_2 \Sigma_\sigma^2 = \sigma + G_\sigma |_{[0,1] \times \{1\}} - G_\sigma |_{\{1\} \times [0,1]} - G_\sigma |_{[0,1] \times \{0\}} .$$

Note that if $\sigma(0) \in V$ then we may assume that $\sigma(0) = \Sigma_0(0)$ and hence $G_\sigma |_{[0,1] \times \{0\}}$ is constant and thus belongs to im ∂_2. The same is true when 1 replaces 0. Moreover if $0 \le c \le 1$ and $e \in E$ then

$$\Sigma(x,y) = e(x + c(1 - x - y))$$

defines a continuous map $\Sigma : \Delta \to X$ such that

$$\partial_2 \Sigma = \Sigma \circ \sigma_1 + \Sigma \circ \sigma_2 + \Sigma \circ \sigma_3$$

where

$$\Sigma \circ \sigma_1(t) = e(c + (1-c)t), \quad \Sigma \circ \sigma_2(t) = e(1-t), \quad \Sigma \circ \sigma_3(t) = e(ct).$$

Using this it is easy to check firstly using the case $c = 0$ that

$$e + e \text{ or } r \in \text{im } \partial_2$$

for any $e \in E$ and secondly by induction on k using $e = G_\sigma \mid_{\{1\} \times [0,1]}$ and $c = t_{k-1}$ that

$$G_\sigma \mid_{\{1\} \times [0,1]} \in C_1^T(X) + \text{im } \partial_2$$

because

$$G_\sigma(1, t) = \Sigma_i \left(\frac{t - t_{i-1}}{t_i - t_{i-1}} \right) \quad \text{if } t \in [t_{i-1}, t_i]$$

for $1 \leq i \leq k$.

Finally suppose that $\xi \in C_1(X)$ and $\partial_1 \xi \in C_0^T(X)$. Then

$$\xi = \sum \lambda_\sigma \sigma$$

where the sum is over all $\sigma \in \text{Map }([0,1], X)$ and only finitely many λ_σ are nonzero. Since $\partial_1 \xi \in C_0^T(X)$ we can choose G_σ as above for each σ with $\lambda_\sigma \neq 0$ in such a way that

$$\sum \lambda_\sigma (G_\sigma \mid_{[0,1] \times \{1\}} - G_\sigma \mid_{[0,1] \times \{0\}})$$

is a sum of constant maps in $C_1(X)$ and hence lies in im ∂_2. Thus

$$\xi = \sum \lambda_\sigma \sigma \in \sum \lambda_\sigma G_\sigma \mid_{\{1\} \times [0,1]} + \text{im } \partial_2 \subseteq C_1^T(X) + \text{im } \partial_2.$$

This proves lemma C.14. $\square$

The proof of lemma C.15 uses just the same ideas as went into the proof of lemma C.14 though the notation becomes more cumbersome, so it will be omitted.

C.3   Spheres with handles

In this section it will be shown that any compact connected surface with a triangulation in the sense of definition 4.9 is homeomorphic to a sphere with g handles, for some $g \geq 0$. We shall follow the argument of [Springer 57] §5.5.

So suppose that (V, E, F) is a triangulation on a compact surface S satisfying 4.9(i)–(v). First we will "flatten out" S using the triangulation and show that it is homeomorphic to a polygon in the Euclidean plane with its edges identified in pairs. This is done as follows.

First choose any face $f_1 : \Delta \to S$ in F. Then choose another face $f_2 : \Delta \to S$ in F with an edge $e_1 : [0,1] \to C$ in common with f_1, that is,

$$e_1 = e_{f_1}^i = e_{f_2}^j$$

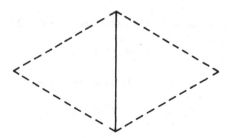

Figure C.2: $\Delta_1 \cup \Delta_2$

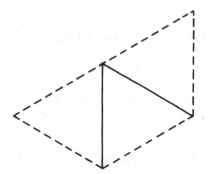

Figure C.3: $\Delta_1 \cup \Delta_2 \cup \Delta_3$

for some $i,j \in \{1,2,3\}$. Then f_1 and f_2 can be used to construct a homeomorphism from the interior of the union of two triangles $\Delta_1 \cup \Delta_2$ in $\mathbf{R}^2$ intersecting along a common edge (figure C.2)to

$$f_1(\Delta^0) \cup e_1((0,1)) \cup f_2(\Delta^0).$$

Now choose another face $f_3 : \Delta \to C$ in F with an edge $e_2 : [0,1] \to C$ in common with f_1 or f_2. Then f_1, f_2 and f_3 can be used to construct a homeomorphism to

$$f_1(\Delta^0) \cup f_2(\Delta^0) \cup f_3(\Delta^0) \cup e_1((0,1)) \cup e_2((0,1)),$$

from the interior of $\Delta_1 \cup \Delta_2 \cup \Delta_3$ where Δ_3 is a third triangle in $\mathbf{R}^2$ intersecting $\Delta_1 \cup \Delta_2$ along an appropriate edge of Δ_1 or Δ_2 (figure C.3). Moreover $\Delta_1 \cup \Delta_2 \cup \Delta_3$ is homeomorphic to a regular pentagon in $\mathbf{R}^2$, so we can replace $\Delta_1 \cup \Delta_2 \cup \Delta_3$ by a regular pentagon if we wish.

We can continue this process until all the faces in F have been used up, since S is connected. For if at some stage we still had some faces unused, but none of them had an edge in common with any face already used, then the unions A and B of the images in S of the faces already used and those not yet

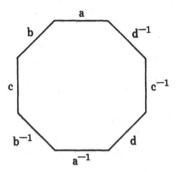

Figure C.4: $abcb^{-1}a^{-1}dc^{-1}d^{-1}$

used would be disjoint closed nonempty subsets of S such that $A \cup B = S$, and this would contradict the connectedness of S.

At the end of this process we obtain a polygon Π in $\mathbf{R}^2$ and a homeomorphism h from the interior Π^0 of Π to S such that h extends to a surjective continous map $h : \Pi \to S$ which by 4.9(v) identifies edges of Π in pairs. Thus topologically S is obtained from Π by glueing together edges of Π in pairs. Moreover it follows from 4.9(v) that a pair of edges of Π must be glued together using opposite orientations: if the first edge is regarded as running anticlockwise around Π then it is glued to the second edge running clockwise.

The precise situation can be symbolised as follows. We label each pair of edges of Π to be glued together as a and a^{-1}, b and b^{-1}, c and c^{-1} and so on, and then write down these labels in the order in which they are encountered when travelling anticlockwise around the boundary of Π. The result is called a *symbol* for the map $h : \Pi \to S$, or simply a symbol for S. For example,

$$abcb^{-1}a^{-1}dc^{-1}d^{-1}$$

symbolises figure C.4 or equivalently figure C.5. We write $(a^{-1})^{-1} = a$ and so on. Of course we can cyclically permute any symbol for S and get another symbol for S. We can also cut the polygon Π into two polygons Π_1 and Π_2 along any line joining two of its vertices, and then attach Π_1 and Π_2 along a pair of identified edges, to obtain a new polygon representing S and hence a new symbol for S. Using this process we can prove that S always has a symbol of a particularly simple form[1].

Theorem C.16 S has a symbol of the form
(i) aa^{-1}, or

---

[1]Since we want to allow symbols of length 2 we must allow polygons with two vertices and curved edges. All polygons with at least three vertices may be assumed to have straight edges as usual.

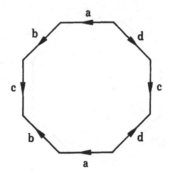

Figure C.5: $abcb^{-1}a^{-1}dc^{-1}d^{-1}$

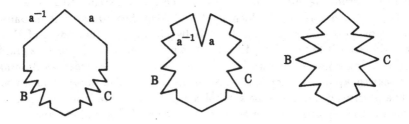

Figure C.6: Removing aa^{-1}

(ii) $a_1b_1a_1^{-1}b_1^{-1}a_2b_2a_2^{-1}b_2^{-1}\ldots a_gb_ga_g^{-1}b_g^{-1}$
for some $g \geq 1$.

Proof. We know that S has a symbol. We shall show that we can modify this symbol to obtain another symbol for S of the form (i) or (ii).

First note that if the sequence aa^{-1} appears in the symbol and if the symbol has at least one other letter then the letters aa^{-1} can be removed from the symbol to obtain a new symbol for S (see figure C.6). Thus either our symbol is of the form (i), or we may assume (as we shall henceforth) that it contains no sequences of the form aa^{-1}.

Now we can assume that all the vertices of the polygon Π correspond to the same point of S. For suppose that there is an edge of Π labelled by a whose vertices P and Q correspond to different points p and q of S. We shall show how to find a polygon representing S with the same number of vertices as Π but one more vertex corresponding to the point p of S; then repeating

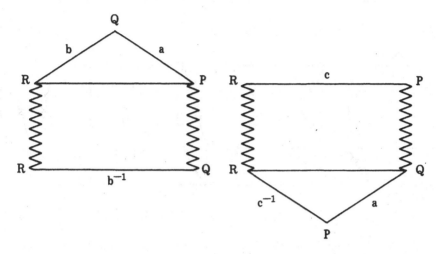

Figure C.7: Increasing the number of vertices corresponding to p

this process enough times will lead to the required situation. We let b be the other edge of Π with Q as a vertex, and join the other vertex, R say, of b to the vertex P of a by a diagonal c of Π to form a triangle with edges a, b, c. Note that b is not a^{-1} since we are assuming that the symbol contains no sequences of the form aa^{-1}. We now cut Π along c and glue back the triangle with edges a, b, c to the rest of Π by attaching the edge b of the triangle to the edge b^{-1} of the rest of Π (see figure C.7). This increases the number of vertices corresponding to the point p by one as required.

Now let us call a pair of edges a and b *linked* if the edges a, b, a^{-1}, b^{-1} occur in the symbol in that order (up to cyclic permutations) but not necessarily next to each other. Then each edge a is linked to some other edge b. For if not then whenever any b occurs between a and a^{-1} in the symbol then b^{-1} occurs between a and a^{-1} too. This means that if we choose a point on a other than a vertex then the line segment joining this point to the equivalent point on a^{-1} divides Π into two polygons Π_1 and Π_2 such that the *only* points of Π_1 which get identified in S to points of Π_2 are those along this line segment and the vertices of Π (all of which correspond to the same point $p \in S$ by our last assumption). This tells us that the complement of the image of this line segment in S is an open subset of S which is connected but which becomes disconnected if the point p is removed from it. It is not difficult to see that no surface S can have an open subset with this property.

Finally we can apply the process indicated in figure C.8 to each linked pair a and b of edges to get a symbol for S in which a, b, a^{-1}, b^{-1} occur successively for each linked pair a and b. This completes the proof of the theorem. $\square$

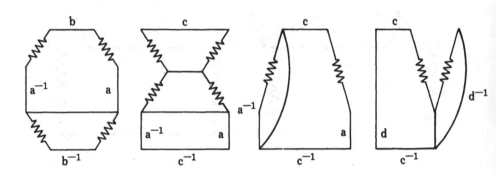

Figure C.8: Modifying linked pairs

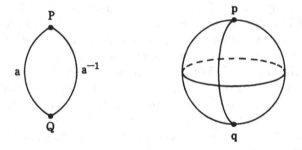

Figure C.9: aa^{-1}

Theorem C.17 *Let S be a compact connected surface with a triangulation in the sense of 4.9. Then S is homeomorphic to a sphere with g handles for some $g \geq 0$.*

Proof. By theorem C.16 S is homeomorphic to a polygon Π in $\mathbf{R}^2$ with its edges identified in pairs according to the symbol

$$aa^{-1}$$

or

$$a_1 b_1 a_1^{-1} b_1^{-1} a_2 b_2 a_2^{-1} b_2^{-1} \ldots a_g b_g a_g^{-1} b_g^{-1}$$

for some $g \geq 1$. In the first case (see figure C.9) S is homeomorphic to a sphere (with no handles). In the second case when $g = 1$ the polygon Π is a

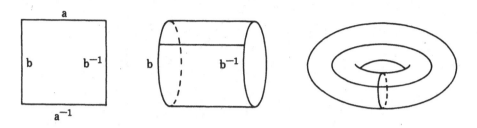

Figure C.10: $aba^{-1}b^{-1}$

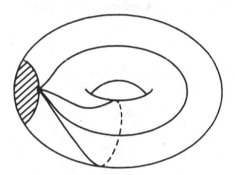

Figure C.11: A torus with a disc removed

square with the opposite pairs of sides identified. Glueing together one pair gives us a cylinder, and then glueing together the ends of the cylinder gives us a torus (see figure C.10).

To see why a torus is topologically a sphere with one handle, cut out a small disc from the torus as in figure C.11 to leave a handle. The small disc which was removed is homeomorphic to the complement of a small disc in a sphere. Glueing the handle back to this gives us a representation of the torus as a sphere with one handle (see figure C.12).

We can assume that the boundary of the disc cut out of the torus passes through the point corresponding to the vertices of the square Π (see figure C.11). Thus cutting the disc out of the torus corresponds to cutting a disc whose boundary contains a vertex out of the square Π (see figure C.13).
Thus we can represent the handle by a pentagon with sides identified as in figure C.14.

Now consider a polygon Π with $4g$ sides to be identified according to the symbol

$$a_1 b_1 a_1^{-1} b_1^{-1} a_2 b_2 a_2^{-1} b_2^{-1} \ldots a_g b_g a_g^{-1} b_g^{-1}.$$

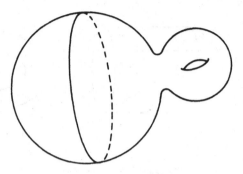

Figure C.12: A sphere with one handle

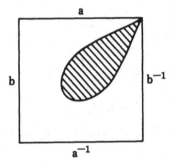

Figure C.13: A cut torus with a disc removed

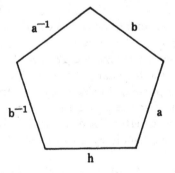

Figure C.14: A handle

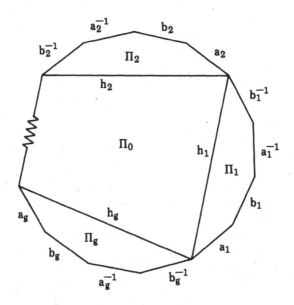

Figure C.15: A subdivision of Π

We can subdivide Π into $g+1$ polygons Π_0 and $\Pi_1, \ldots, \Pi_g$ as indicated in figure C.15 by joining the vertex between b_{j-1}^{-1} and a_j to the vertex between b_j^{-1} and a_{j+1} for each $j \in \{1, \ldots, g\}$ modulo g. If $1 \leq j \leq g$ the polygon Π_g represents a handle with boundary curve h_j. The polygon Π_0 has all its vertices identified, and hence represents a sphere with g discs cut out of it, where the boundaries of the discs (which are the curves $h_1, \ldots, h_g$) all meet at a point but otherwise the discs do not intersect (see figure C.16).

Thus the polygon Π represents this sphere with a handle attached to the boundary of each of the removed discs. The result follows. □

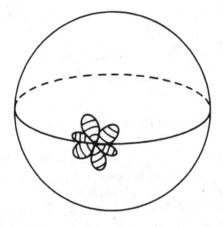

Figure C.16: The space represented by Π_0

Bibliography

[Arbarello & al 85]   E. Arbarello, M.Cornalba, P.Griffiths and J. Harris, *Topics in the theory of algebraic curves*, Springer-Verlag (1985).

[Arnol'd 86]   V.I. Arnol'd, *Catastrophe theory*, Second Edition, Springer-Verlag (1986).

[Atiyah & Macdonald 69]   M.F. Atiyah and I.G. Macdonald, *Commutative algebra*, Addison-Wesley (1969).

[Beardon 84]   A.F. Beardon, *A primer on Riemann surfaces*, London Math. Soc. Lecture Notes 78, Cambridge (1984).

[Brieskorn & Knörrer 86]   E.Brieskorn and H.Knörrer, *Plane algebraic curves*, Birkhäuser-Verlag (1986).

[Chern 80]   S.S. Chern, *Complex manifolds without potential theory*, Springer-Verlag (1980).

[Clemens 80]   C.H. Clemens, *A scrapbook of complex curve theory*, Plenum (1980).

[Coolidge 59]   J.L. Coolidge, *A treatise on algebraic plane curves*, Dover (1959).

[Farkas & Kra 80]   H. Farkas and I. Kra, *Riemann surfaces*, Springer-Verlag (1980).

[Fulton 69]   W. Fulton, *Algebraic curves*, Benjamin- Cummings (1969).

[Griffiths 89]   P.A. Griffiths, *Introduction to algebraic curves*, Transactions of mathematical monographs 76, American Mathematical Society (1989).

[Griffiths & Harris 78]   P.A. Griffiths and J. Harris, *Principles of algebraic geometry*, Wiley (1978).

[Gunning 66]   R.C. Gunning, *Lectures on Riemann surfaces*, Princeton (1966).

[Hartshorne 77]   R. Hartshorne, *Algebraic geometry*, Springer-Verlag (1977).

[Herstein 75]   I.N. Herstein, *Topics in algebra*, Wiley (1975).

[Jones 71]   B.F. Jones, *Rudiments of Riemann surfaces*, Rice University (1971).

[Kendig 77]   K.Kendig, *Elementary algebraic geometry*, Springer-Verlag (1977).

[Morrow & Kodaira 71]   J. Morrow and K. Kodaira, *Complex manifolds*, Holt Rinehart & Winston (1971).

[Mumford 75]   D. Mumford, *Curves and their Jacobians*, University of Michigan (1975).

[Mumford 76]   D. Mumford, *Algebraic geometry I: Complex projective varieties*, Springer-Verlag (1976).

[Priestley 85]   H.A. Priestley, *Introduction to complex analysis*, Oxford (1985).

[Reid 88]   M. Reid, *Undergraduate algebraic geometry*, London Math. Soc. Student Texts 12, Cambridge (1988).

[Semple & Roth 49]   J.G. Semple and L. Roth, *Introduction to algebraic geometry*, Oxford (1949).

[Shafarevich 74]   I.R. Shafarevich, *Basic algebraic geometry*, Springer-Verlag (1974).

[Spanier 66]   E.H. Spanier, *Algebraic topology*, McGraw-Hill (1966).

[Springer 57]   G. Springer, *Introduction to Riemann surfaces*, Addison-Wesley (1957).

[Stewart 73]   I. Stewart, *Galois theory*, Chapman & Hall (1973).

[Sutherland 75]   W.A. Sutherland, *Introduction to metric and topological spaces*, Oxford (1975).

[Walker 50]   R.J. Walker, *Algebraic curves*, Princeton (1950).

Index